Practical Electric Motor Handb

Practical Electric Motor Handbook

Irving M. Gottlieb PE

Newnes
An imprint of Butterworth-Heinemann
Linacre House, Jordan Hill, Oxford OX2 8DP
A division of Reed Educational and Professional Publishing Ltd

A member of the Reed Elsevier plc group

OXFORD BOSTON JOHANNESBURG
MELBOURNE NEW DELHI SINGAPORE

First published 1997

© Irving M. Gottlieb 1997

All rights reserved. No part of this publication may be reproduced in any material form (including photocopying or storing in any medium by electronic means and whether or not transiently or incidentally to some other use of this publication) without the written permission of the copyright holder except in accordance with the provisions of the Copyright, Designs and Patents Act 1988 or under the terms of a licence issued by the Copyright Licensing Agency Ltd, 90 Tottenham Court Rd, London, England W1P 9HE. Applications for the copyright holder's written permission to reproduce any part of this publication should be addressed to the publishers.

British Library Cataloguing in Publication Data

A catalogue record for this book is available from the British Library

ISBN 0 7506 3638 2

Library of Congress Cataloguing in Publication Data

A catalogue record for this book is available from the Library of Congress

Typeset by Vision Typesetting, Manchester
Printed and bound in Great Britain by Biddles Ltd, Guildford and King's Lynn

Contents

Preface	ix
1 Electric motor generalities	**1**
Early discoveries	2
The quest for continuous rotary motion	3
Basic motor action	5
The electric motor as an energy converter	7
Motor graphs	8
Motor nomenclature	9
Horsepower rating of electric motors	11
Motor classification	12
Describing performance of electric motors	13
Illustrations pertaining to motors	13
Measuring speed	15
Applying 'hand-rules' to motors	17
Force, current and flux – the orthogonal relationship	18
Counter-e.m.f.	20
Electric motors in the quest for perpetual motion	22
Idealized concept of energy conversion	23
Measurement pitfalls	26
The electric vehicle	26
Things to keep in mind about motors	31
2 Practical aspects of DC motors	**32**
Background of DC motors	32
The homopolar motor	33
The AC involvement in DC motors	34
A practical view of armature reaction	36
The role of residual magnetism	38
The DC shunt motor	39
The DC permanent magnet motor	42

Paradoxes in DC motor theory	44
Operation of the DC servo motor	47
The motor characteristic of the DC watthour meter	49
The DC series motor	51
The DC compound motor	53
Motor and generator performance in the same machine	54
Reversing the rotation of DC motors	56
Practical use of counter-e.m.f.	57
Flexible control of permanent magnet motors	59
The grey area of DC and AC motors	60

3 Practical aspects of AC motors — **62**

The great induction motor dilemma	63
Practical aspects of the single-phase induction motor	63
Split-phase starting techniques for induction motors	65
Types of single-phase AC motors	66
The synchronous motor	68
Shaded-pole motors	72
The hysteresis motor	73
The reluctance motor	75
The wound-rotor induction motor	76
The double squirrel-cage induction motor	80
Speed control of AC motors	82
The consequent-pole AC motor	82
Speed selection by pole modification	84
The universal motor	87
Rotation reversal in AC motors	88
Non-sinusoidal waveforms applied to AC motors	91
Power input and power factor in three-phase induction motors	94
Unusual motor behaviour	96

4 Practical projects — **99**

Experimental aspects of electric motors	99
Considerations in starting electric motors	99
High starting-torque from a small capacitor	101
Gentle start-up of AC motors	103
An easy starting technique for three-phase induction motors	104
Altering the characteristics of the series motor	105
Phase transformation for motor testing	107
Digitally-generated polyphase waveforms	108
Synchro-system experiments	109
Operating AC motors from 50 Hz or 60 Hz lines	112
Changing the function of a dynamo	113
Dynamic braking of permanent magnet motors	115

Dynamic braking of induction motors	115
Speed control of fan motors	118
Speed control with and without speed-regulation	118
Practical aspects of the brushless DC motor	120
Improving performance of stepping motors	122
Custom-designing of stepping motors	122
Electronic technique for eliminating the centrifugal switch	124
Some motor drive techniques	124
IC control system for permanent magnet DC motors	128
IC control system for brushless DC motor	130
IC energy-saving system for induction motors	132
DC permanent magnet motor for electric vehicles	133
Switched reluctance motor	135
Reliability	136

5 Practical problems — 138

Dealing with motor mathematics	138
Feeder line as part of the motor circuit	138
Internal power in a permanent magnet DC motor	139
Stray power in a permanent magnet DC motor	140
Power determinations from prony-brake measurements	141
Compound motor speed calculations	142
Speed regulation of DC and AC motors	144
Torque calculation of DC series motor	145
Efficiency of DC shunt motor	145
Starting resistance for permanent magnet DC motor	146
Power requirement from electric vehicle motor	147
Speed vs. load for DC shunt motor	148
Speed vs. load for DC series motor	149
Synchronous speed calculations	150
Light-load behaviour of induction motor	151
Using wattmeter data for evaluating induction motor performance	152
Speed of wound-rotor induction motor vs. rotor resistance	153
Using 50 Hz motors on 60 Hz lines and vice-versa	154
Interpreting data for polyphase motors	154
Capacitor calculation for unity power factor	155
Synchronous motor calculations for power factor correction	156
Transformer simulation of induction motor	157
Two-wattmeter data for induction motor calculations	158
Calculations for induction–generator action from a motor	159

Index   163

Preface

Many good books are available which provide a rigorous and comprehensive treatment of electric motors. These serve the needs of academia, and are fine for both would-be and accomplished specialists. There are, however, numerous technologists and practitioners of the applied sciences who may not readily derive benefit from such treatises; for instance, engineers, electronics designers, intelligent hobbyists and experimenters. Although such people generally possess more-than-adequate technical backgrounds, they often feel ill at ease when working with electric motors. Included in their company are electrical engineers for the simple reason that their training probably focused more on software, programming and computer logic than on rotating machinery.

This book therefore targets the large body of workers reasonably versed in engineering concepts who feel the need of practical insights relating to electric motors. Rather than motor design, their chief concerns lie with the selection, system installation, operation and performance evaluation of electric motors. In the pursuit of this goal, the author has sought to clarify those aspects of electric motors that all too often pose difficulties for both students and professionals. Electronic specialists with expertise in analog and digital control techniques should recognize many possibilities of modifying the 'natural' characteristics of electric motors. Even those interested in the detailed nuances of specialized design, should find useful guidance in this practical treatment of electric motors.

1 Electric motor generalities

Historians like to assign definite dates to mark the occurrence of significant events. This is not quite so easy to do in science and technology as it is in, say, politics. When one studies the birth and evolution of notable achievements in either theoretical or applied science a great deal of fuzzy logic is encountered in attempts to date the sudden emergence of the event, and more 'originators', inventors, discoverers and improvers are usually involved than given deserved credit. Moreover, there are inevitably earlier workers in the field who laid down the basic intellectual tools for demonstrable ideas and devices.

This has been true for electric motors, as well as for aircraft, telephones, incandescent lamps, internal combustion engines, etc. Indeed, near or actual simultaneous invention has been the order of the day – it is as if thought patterns and variations of previous ideas are forever 'in the air'.

It is fitting, therefore, to at least recall the names of several of those who can be said to be the more-or-less immediate precursors of the electric motor. In 1819, Hans Christian Oersted noted the physical deflection of a magnetized needle near a current-carrying conductor (See Fig. 1.1). Shortly after, Michael Faraday successfully produced continuous rotary motion in an otherwise impractical electric device. Later, he devised the very practical Faraday disc, which could perform as either a generator or a motor. Joseph Henry, a near-contemporary of Faraday, did pioneering work in laying down the rules of electromagnetic induction. The overlap between the experimentation of Faraday and Henry bears witness to the alluded 'ideas in the air'.

Lenz's law, propounded by Heinrich Lenz in 1833, also contributed heavily to electric motor technology. His rule – that the mechanical action involved in inducing electric current is opposed by the resultant magnetic field – affects both the design and operation of electric motors. Science

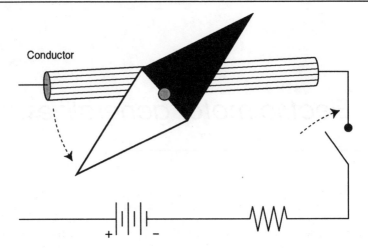

Fig. 1.1 One of the earliest indications of motor action. To the alert mind, primitive experiments can reveal the possibility of practical devices. The above set-up replicates the observation of Hans Christian Oersted that a freely pivoted magnetized needle (or compass) can undergo a physical deflection in the presence of a current-carrying conductor. Study and contemplation of this phenomenon led to understanding of the all-important interactions involving electricity, magnetism and mechanical force or physical motion.

history can, of course, be telescoped backwards to ancient times, but these pioneers were notably active in ushering in our modern era.

Early discoveries

Although the conversion of electricity into mechanical motion has become a mundane expression of familiar hardware, neither physics nor mathematics provide completely satisfying explanations of the involved phenomena. It is easy enough to recite, parrot fashion, textbook statements that magnets can attract or repel one another, that a current-carrying conductor is encircled by magnetic lines of force, etc. Yet the very notion of provoked action at a distance entails a hidden mystery. Nature reveals force fields that exert influence on bodies and on other fields; neither a vacuum nor astronomical distances constitute barriers to these actions and interactions. Although we learn to accept the reality of action at a distance, it can still instill in us a sense of mystery.

Gravity, electrostatics and the nuclear force are tantalizingly suggestive of at least some of the attributes of magnetism. It is the differences that are hard to understand. For example, how can we make a gravitational motor? Capturing some kinetic energy from a waterfall could be offered as an answer, but we would really like to directly manipulate gravity somewhat as

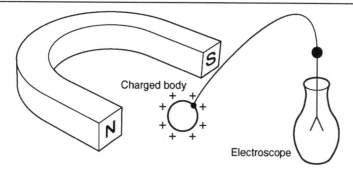

Fig. 1.2 *The electricity–magnetism link eluded early experimenters. The ultimate discovery of the interaction between the two manifestations of nature was the precursor of electric-motor technology. An experiment such as that shown suggested independent and isolated existences for electricity and magnetism inasmuch as nothing was observed to happen. We are similarly frustrated today in our inability to prove where gravitational force fits into the scheme of things.*

we manipulate magnetism. And no repulsive gravitational fields have been found that would make levitation possible. Moreover, if we didn't already know how to pursue the matter further, it could be easily concluded that magnetic and electric fields lead isolated existences devoid of possible interactions. For instance, a charged particle situated between the poles of a horseshoe magnet does nothing at all; nor does the magnetic flux pay any heed to the stationary charged particle. Figure 1.2 replicates such an experiment.

When the scientists and experimentalists of the nineteenth century observed the reversible relationship between *moving* electric charges and magnetism, they quickly made another fortuitous discovery – it was found that a third parameter was associated with this linkage. This was *physical motion*. That is, a current-carrying conductor in a magnetic field could experience motion. And, in harmony with a symmetry often seen in nature, a moving conductor in a magnetic field developed a voltage across its ends. Because these unexpected interactions were duly noted, the birth of electric motors (and generators) was ensured.

The quest for continuous rotary motion

From our present vantage point, the chance observation that a magnetized needle was deflected by a current-carrying conductor appears a triviality scarcely worthy of mention. Yet the application of such a cause-and-effect relationship to continuous rotation must have tantalized the curious minds of the day. It is to be recalled that many manifestations of electricity and magnetism had been recognized for centuries, but the utilization of a force

4   *Practical Electric Motor Handbook*

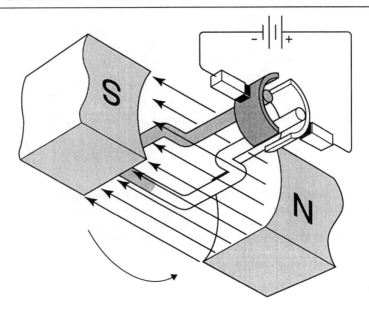

Fig. 1.3 *The basic DC electric motor.* Continuous rotation is the salient feature of this set-up. Unidirectional development of electromagnetic torque takes place due to the current-reversing action of the brush–commutator system. The principles underlying the operation of the toy-like assembly of elements depicted above are basic to design of practical electric motors.

derived from linkage of the two entities somehow eluded all who 'played' with them.

Once, however, production of a physical force was noted, the problem of translation into continuous rotary motion intrigued the advanced experimenters. One solution, the Faraday disc, proved that it could be done. However, the extremes of high current and low voltage made this motor difficult to use in the practical world. A much more practical DC motor emerged in which a mechanically driven switch timed the current flow in conductors in such a way as to always subject them to unidirectional torque in the presence of a magnetic field. Thus, was born the brush and commutator system giving rise to the practical electric motors needed by the budding industrial age.

From even a toy-like model of a primitive commutator-type DC motor, such as illustrated in Fig. 1.3, the following useful information can be gleaned:

(1) The polarity of the DC source determines the direction of rotation.
(2) Maximum electromagnetic torque occurs with the rotating element, i.e., the armature, in the position shown. Conversely, zero torque exists in the position depicted in Fig. 1.4.

Electric motor generalities 5

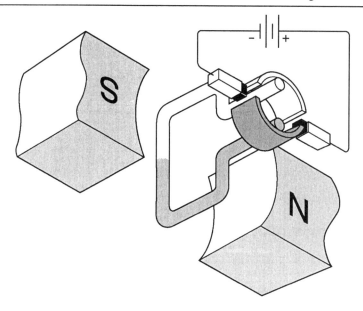

Fig. 1.4 *Zero-torque position of the armature conductors. The primitive motor with a single armature-loop delivers a pulsating torque. It cannot start if positioned as illustrated at standstill. The remedy in practical motors is to provide multiple loops spaced so that one or more is always in a torque-generating position. Practical motors also have multiple-segment commutators.*

(3) The magnetic power is not 'used-up' by the operation of the motor.
(4) Increasing the field strength from the magnet and/or the current supplied, increases the mechanical power available from the shaft.
(5) Alternating current flows in the armature when the motor is operating.
(6) Notwithstanding the revelation of (5), the motor cannot operate from an alternating-current source.

Basic motor action

The magnetic field surrounding a current-carrying conductor figures prominently in the interactions giving rise to basic motor action. The simple experiment shown in Fig. 1.5 demonstrates the concentric pattern, as well as the directivity of the current produced flux. Readers familiar with the practicalities of toroids, solenoids, inductors, transformers, etc. may recall rather uninteresting expositions of this topic in their training texts. The point to be made here is that this concentric flux around a current-carrying conductor lies at the very heart of the force manifested as 'motor action'. How this comes about may be gleaned from the situation depicted in Fig. 1.6.

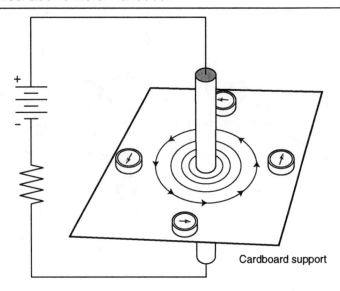

Fig. 1.5 *Concentric magnetic flux around a current-carrying conductor. Either several compasses, or a single compass moved in successive positions around the conductor will serve the purpose of the experiment. The circular pattern of the magnetic field plays a prominent role in the armatures, field windings, stators, and rotors of the various types of electric motors. Significantly in motor operation, a reversal in current direction reverses the direction of the magnetic lines of force.*

Here we see a current-carrying conductor immersed in a magnetic field provided by the poles of a horseshoe magnet. The net field due to the interaction of the circular field of the conductor and the otherwise-linear field from the poles of the magnet are greatly distorted. One can visualize the resemblance of this magnetic flux pattern with the pressure inequalities causing the lift of an aircraft wing. In any event, it is evident that there is dense magnetic flux on the bottom surface of the conductor and sparse flux on the top. Not only do the magnetic lines of force constituting the flux display rubber-band physical properties, but they strongly repel one another. It is thus easily seen that this distorted field pattern must exert an upward force on the current-carrying conductor. We have, in other words, 'motor action'. Note that a reversal of *either* the direction of the main field from the magnet, or the direction of the current in the conductor will produce *downward* motor action.

Besides the physical motion of the current-carrying conductor in Fig. 1.6, or more precisely, *because of it*, a voltage is induced in the conductor so polarized as to oppose the current causing the motor action. This simultaneous behaviour as a *generator* is the practical manifestation of Lenz's law. In a

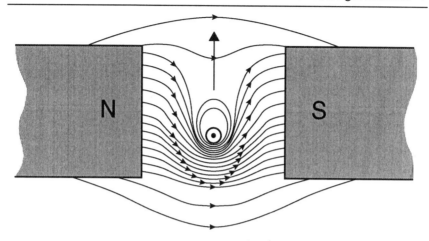

Fig. 1.6 *Motor action exerted on current-carrying conductor in a magnetic field. Endowing magnetic lines of force with the elastic property of rubberbands, enables one to visualize the motion imparted to a current-carrying conductor. The interaction of the magnetic fields as shown is found in virtually all electric motors. Downward motion of the conductor would occur if either (not both) the current direction or the magnetic poles were reversed. Note: Conventional current-flow is used in this book.*

general, but inviolate way, it tells us that 'any change in magnetic flux linkage is accompanied by effects *opposing* the change'.

The electric motor as an energy converter

At the very outset, we should concern ourselves with what electric motors do. A popular but erroneous notion is that electric motors create or produce mechanical energy. Mechanical energy is definitely not *created*; yes, it may be said to be *produced* at the shaft of the motor, but this is, at best, only a partial answer. We must point out that this mechanical energy comes at the *expense* of some other form of energy. The simple and true fact of the matter is that the electric motor (and the electric generator, as well) is an *energy converter*. More specifically, the motor converts electrical energy into mechanical energy. In so doing, it is never 100% efficient – in the overall budget of energy availability, there are always inevitable energy losses. These losses may manifest themselves as still other forms of energy, such as heat, light, sound, friction, radiation, etc.

Energy, itself is the capability of doing work. In the practical world, it would be well to say that *available* energy represents the capability of doing *useful* work. Because of nature's previous activities, most of the useful energy sources stem from various chemical, gravitational, and nuclear arrangements of planetary matter. In contrast to such earthly energy sources, solar radiation represents a dynamic and ongoing source of energy. All our electric motor does or can do is to directly or indirectly participate as an energy converter in which another form(s) of energy gets transformed into our desired mechanical energy. Practically, we see this conversion or transformation as electricity *in* and mechanical work *out*.

Power and energy tend to be used interchangeably in popular communications. Power is the *rate* of energy transfer. Or in other words, energy is the *product* of power and *time*. Thus, our monthly utility bill is based upon a number of kilowatt-hours.

We, on earth can transform energy, but cannot create it. Interestingly, those seeking to circumvent natural law seem 'magnetically' attracted to electric motors. Such claims as the following routinely litter the desks of patent clerks and editors.

Motor graphs

Many graphs depicting motor performance show some parameter as a function of the line current or armature current, these being virtually the same quantity. For example, one might see speed or torque as the ordinate (the vertical axis) of the graph plotted against armature or line current as the abscissa (the horizontal axis of the graph). One naturally infers that the armature current is *somehow* varied and the corresponding values of speed or torque are then either measured or calculated. Those not familiar with motors usually suppose that the armature current is adjusted by means of a rheostat, a variable auto-transformer, or an adjustable power supply. This is *not* the case. Refer to Fig. 1.7.

The key word above is 'somehow'. The actual situation is that the armature current is *caused* to vary by applying different mechanical loads to the motor. In other words, the armature current reflects load changes. It is true that it would be difficult to determine the actual load values; armature current tracks load changes and is very easy to observe with an ammeter inserted in the motor line. Moreover, direct manipulation of the current would introduce complications in the interpretation of the results. Reiterating, a *variable load* is used to plot the majority of these graphs. This practice is so universal that it is often not explained that the various motor currents used to plot the graph are due to variation in the load applied to the shaft of the motor. It is simply assumed this is common knowledge, and often, it is a stumbling block for students.

On the other hand, it should not be assumed that the direct electrical variation of armature or line current is not a permissible and useful technique for certain applications. *Here*, however, the wise practitioner would append a notice to a graph showing the speed or torque relationship to

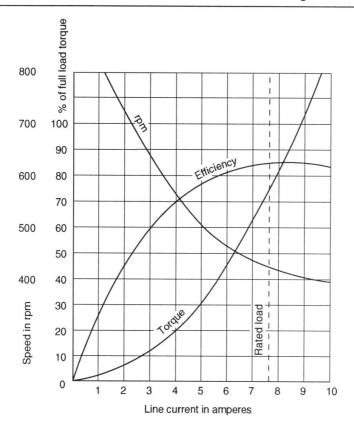

Fig. 1.7 *Graphical representation of the characteristics of a DC series-motor. A typical graph such as this could be misleading to persons not familiar with electric motor technology. The line current is not varied by a rheostat, autotransformer, or by any other means. Rather, the mechanical load imposed on the motor is varied and the corresponding line currents are recorded and plotted on the horizontal axis of the graph. This would, no doubt be clearer if the caption read 'Line Current in Amperes Due to Load'.*

armature or line current, stipulating that the relationships were valid under the condition of *constant load*.

Motor nomenclature

Initial exposure to some of the nomenclature pertaining to electric motors can be confusing. An *armature*, to be sure, is the rotating member of DC motors. It is also the stationary member of certain AC motors. See Fig. 1.8. Although the physical difference is obvious, the identity of their electrical functions is not altogether a clear issue. Moreover, the field-winding of

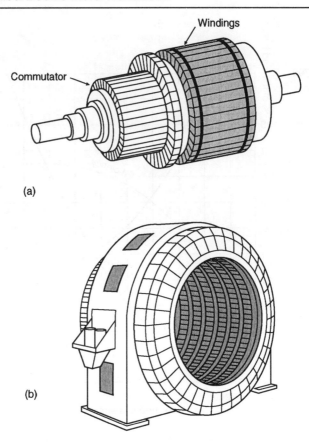

Fig. 1.8 *Armatures of entirely different dynamos. (a) The armature of a DC motor. (Also similar to those used in AC repulsion motors.) (b) The armature of an AC three-phase induction motor. Confusion can be avoided by referring to the stationary winding of AC motors and alternators as the* stator.

motors can be found as either the rotating or the stationary member. It follows that the same can be said for permanent-magnet fields. The overall situation is not clarified by allusion to rotating fields – these can be developed by physically rotating magnets or electro-magnets, or by stationary armatures impressed with polyphase currents.

Fortunately, such confusion can be resolved by using the term *stator* for the stationary member of all AC motors. Similarly, it is helpful to apply the term *rotor* to the rotating members of these motors. (Stepping motors and DC brushless motors, because they bear some constructional similarities to AC synchronous motors, are also said to have rotors.)

Electric motor generalities 11

It is interesting to contemplate that the *stators* of three-phase induction motors, three-phase synchronous motors and three-phase brushless DC motors can be essentially similar. Indeed, the same machine can serve as either an alternator or a synchronous motor. Additionally, the rotating members referred to as armatures of certain AC repulsion-type motors can closely resemble the armatures used in DC motors. Thus, we *can* have an armature and a stator in the same machine.

Concerning repulsion motors, the inference appears to be that other motors are 'attraction' motors. However, Lenz's law shows that the force of repulsion is at the root of motor action in the classic DC and AC motors. (The purist might argue the stepping motor to be the exception, at least when operating in the stepping mode.)

In the AC induction motor, the rotating field of the stator appears to attract the more slowly rotating rotor conductors. If, however, we think of the stator field as being stationary, the *relative* motion of the rotor is in the *opposite* direction to that of the actual rotating field. Thus, motor action arises from *repulsion* as would be predicted by Lenz's law – induced fields oppose the motion responsible for their production.

Horsepower rating of electric motors

To those with limited experience of working with electric motors, some of the observed conventions must appear just a bit strange. For example, when ordering a motor, one refers to its basic ability for converting electrical to mechanical energy by specifying its *horsepower*. Yet, it will be found that most of the manufacturer's data deal with *torque*. A little contemplation reveals the reason for this.

It turns out that torque, the turning effort, is more fundamental than horsepower which is the *rate* of supplying energy. Horsepower is the product of torque times speed, so that a given horsepower can correspond to a high torque and low speed, or to the converse combination. In practical applications, one is usually specifically interested in knowing the torque and the speed *separately* as they apply to the load on the motor. One should note that speed is very easily measured. Because of these considerations, the graphs of motor performance will either depict torque as the function of some other parameter such as armature current, or alternatively some parameter, such as speed, as a function of torque.

More quantatively, torque itself is the product of the force developed at the rim of a disc, cylinder or wheel times the distance to the centre. Thus, pound-feet is a common unit for this measurement.

A practical manifestation of what has been said is the fact that the horsepower output of a motor at standstill is zero. Even giant motors develop *zero* horsepower at the instant an attempt is made to start them. On the other hand, *torque*, and specifically *starting torque*, tells us what we want to know about starting capability. Indeed, this performance characteristic is one of the primary considerations in motor selection and application.

In a general way, horsepower, because it is specified at a rated speed, motor current and motor voltage (and frequency), can provide guidance in selection of the size of the motor. However, in order to know whether it will serve a particular application, we must ascertain that the right *combination* of speed and torque can be delivered.

Motor classification

Practitioners in the various applied sciences tend to view electric motors as generic devices for converting electrical to mechanical energy. Certainly such a concept is entirely valid but in practice, however, it turns out to be just the tip of the iceberg; the very first prerequisite in grasping the basic framework of electric motor technology is an appreciation of the extensive classification needed to deal with these motors in the practical world.

To begin with, there are direct current (DC) and alternating current (AC) motors. The alternating current types are then subdivided into single-phase and different polyphase designs and, of course, the size or capability of the motor is always an all-important issue. But, the power output doesn't tell us enough; we must also have data pertaining to speed and torque and, speaking of torque, a motor cannot render useful service if it won't start; therefore, specific knowledge about its starting torque is always a matter of priority.

Early in our appraisal of an electric motor, we find that its 'packaging' and constructional features merit deliberation. One can specify waterproof or explosion-proof types, or the motor can be packaged so as to be hermetically-sealed. Ventilation and allowable temperature rise should also not be ignored. A system may require vertical mounting of the motor, or there may be a need for dual output shafts. Torque and speed requirements sometimes mandate integrally-mounted gearboxes. Then, there are the ever-present compromises involving bearing-selection against cost, maintenance and longevity.

As if this isn't sufficient, it is important to know the possible side-effects that may plague an otherwise satisfactory operation. Some types of motors are more prone to generating radio and electromagnetic interference than others. Certain alternating-current motors can upset the supply line with a low power-factor.

Finally, because of solid-state electronics and computer techniques, the classification of electric motors according to function and response has become increasingly complex. Interestingly, however, the diversity of motor-types and control techniques now point the way to a widely-expanded range of useful implementations.

Describing performance of electric motors

Work, energy, power, and torque have definite meanings in physics and engineering, as well as being key words in motor technology. Yet, ordinary and often technical literature uses these basic terms in a sloppy manner. At the very outset, it should be understood that *energy* is the capacity for doing work or the accomplishment of such work. In sharp contrast, *power* expresses a rate of energy expenditure. Power multiplied by the time duration over which the power is applied is the energy expended or consumed. Conversely, energy must be divided by the time the energy accumulates in order to obtain power. We are charged for our use of electrical *energy*. For example, one hundred kilowatt-hours (kWh) of electrical energy could result from 100-hours use of a one kilowatt (kW) heater, or from 200 hours use of 500 watts worth of incandescent lamps. These appliances are rated in terms of *power*. If used over a period of *time*, they consume *energy*. Thus, power and energy should not be casually used on an interchangeable basis. For the work–torque conflict, see Fig. 1.9.

All this begs for a definition of *work*. Work and energy are, from a technical viewpoint the same entity. However, good use of the language does not always permit easy interchangeability. Work results when a directional force moves an object in the same direction. If these directions are not the same, it is the component of the force that moves the object in the same direction that the force is acting that is effective in doing work. The unit of work is the *foot-pound*. (It can also be said that work takes place when a force overcomes a resistance. From physics, it can be proved these apparently-different definitions are one and the same.)

Torque also involves the application of a force, but this time against a pivoted moment-arm so that a *turning tendency* is produced. The magnitude of this turning tendency is expressed as so many *pound-feet* as the force in pounds is multiplied by the length of the moment arm. No actual motion has to take place. And even when rotation does occur, the torque, *itself*, does not do work. Torque multiplied by speed yields power and finally, power exerted over a *time-period* is work or energy. The use of the foot-pound unit for torque is sometimes encountered and is wrong!

Illustrations pertaining to motors

One of the intellectual hurdles to be overcome by those either commencing or renewing their acquaintanceship with electric motors has to do with symbols and schematic diagrams. Although seemingly a triviality, certain practices can lead to confusion. Some of this is brought about by ineffective codes of practice, some by the 'lazy draughtsman' syndrome, and some are locked into place through the force of tradition.

Consider, for example the universally-recognized symbol for a *motor*, as

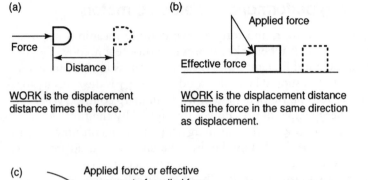

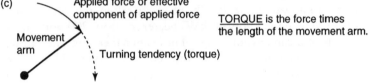

Fig. 1.9 *Foot-pounds and pound-feet: Look-alike units with a difference. Foot-pounds is the unit of work or of mechanical energy. Pound-feet is the torque unit. In motor calculations and specifications, it is important to distinguish between the two entities. Therefore, care is needed not to indiscriminently use these look-alike units.*
(a) Work is represented by displacement of a body in the same direction as the acting force. With force given in pounds and the distance the body is moved given in feet; the unit is foot-pounds.
(b) If the directions of the acting force and the displacement are not the same, only that component of force which acts in the direction of the displacement is effective in producing work.
(c) A force applied perpendicular to a moment-arm produces the turning-tendency called torque. The unit is pound-feet.

depicted in Fig. 1.10(a). Unfortunately, this motor symbol is often used to stand for AC motors that have no commutator–brush system. A better general symbol for an electric motor would be the circle with a capital M, as shown in (b) of Fig. 1.10.

In this book, and in most electric motor literature, the illustrations associated with the theory of AC induction and synchronous motors invariably show stator (armature) structures with salient magnetic poles. Yet, if one of these is examined, one sees no such protruding pole–pieces. Indeed, it is not easy to immediately discern the number and nature of the poles from the *distributed* windings used. When viewing the art, one must think that the operation occurs *as if* the actual machine had these identifiable protruding poles. Also, it should be realized that the depiction is accurate over a small interval of time inasmuch as these poles are either rotating or fluctuating. To

Electric motor generalities 15

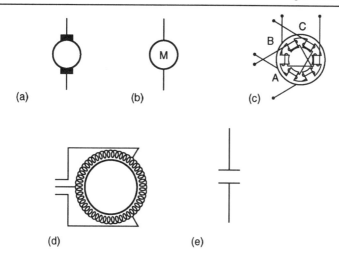

Fig. 1.10 *Some symbology pertaining to electric motors can be confusing.*
(a) This symbol was originally used for DC motors, but has become valid for other motors utilizing brush–commutator systems. It can be confusing to use it for just any motor.
(b) This is a better symbol for AC motors as a class. It also is preferable to (a) when depicting electric motors in general.
(c) Most AC motor stators do not have the salient (protruding) poles commonly used in theory illustrations. Instead, one would see a distributed winding more closely approximated by sketch (d).
(e) Electronics practitioners could initially confuse this sometimes-encountered switch symbol for a capacitor.

make good practical sense, one has to contemplate the art and text *together*.

Also, no matter how many poles may be electrically simulated by such stators, theory is inevitably illustrated by a *two-pole* dynamo. This should be recognized as an artist's short-cut – nothing would be gained from a pictorial drawing of the real machine, or even by laboriously showing a large number of simulated pole-pieces.

Be wary, too, of practices carried over from the electrician's world. Certain switch symbols, for example, could initially be construed to symbolize capacitors by electronics practitioners. The best protection against such confused road-signs is a good measure of practical common sense.

Measuring speed

In the technical literature, one finds reference to the *instantaneous* speed of a synchronous motor. Confusion can arise from such an allusion, for by this implied definition, such motors are supposed to run at the speed of their

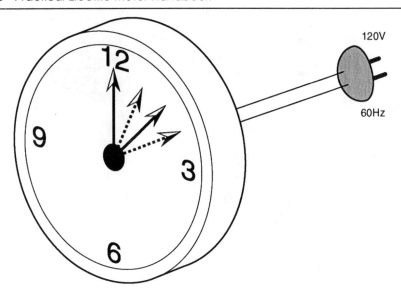

Fig. 1.11 *Greatly exaggerated view of synchronism. Momentarily, the synchronous motor powered clock may speed up and slow down in response to voltage or bearing friction variations. That is, its* instantaneous *speed can fall out of step with the rotating magnetic field which produces the motor's torque. In* most *applications this has no practical consequence because these transient departures from true synchronous speed quickly average out to zero. Thus,* average *speed actually becomes* synchronous *speed over any duration commonly of practical concern. Watch for exceptions, however.*

rotating fields. In turn, the speed of the rotating field is set by the poles and the frequency of the power line. In a given synchronous motor, the speed is determined by the relationship:

$$S = \frac{120(f)}{P}$$

where S represents speed in rpm, f is the powerline frequency in Hz, and P is the number of poles per phase. We know from the use of the small synchronous motor in electric clocks that its speed must be almost a sacred thing. Why then suggest an instantaneous speed? A deviation from perfection is no longer perfection. Fig. 1.11 may help clarify the situation.

There is a bit of Semantic fuzziness here. The *average* speed of such a motor is indeed sacred; if the power-line frequency does not vary at all over a 24-hour period, neither will our clock indicate a false time. Yet, because of possible voltage dips or transients, or even slight load variations from the geartrain in the clock, there certainly will be slight variations in the clock's instantaneous speed. To say that a synchronous motor runs in synchronism with its rotating magnetic field refers to its *average* speed.

Disturbances in operating conditions (other than in frequency) do affect the instantaneous or *momentary* speed. Moreover, such disturbances tend to produce 'hunting' phenomena in which there is damped oscillation of the speed variation. Yet, every speed overshoot will be cancelled by the subsequent undershoot so that average and synchronous speed become one and the same. The non-technical user need not even be aware of these matters.

In contrast, the induction motor tends to approach, but *cannot* attain synchronous speed. Its departure from synchronous speed is known as its *slip*-speed. Sustained slip-speed is a requisite and talk of instantaneous or average speed here is practically meaningless in the sense that these terms apply to synchronous motors.

Applying 'hand rules' to motors

As a result of the lack of perfect standardization in scientific and technological concepts, confusion can set in when studying from several textbooks. A case in point has to do with 'hand rules' for determining the operating parameters of motors. Reference to Fig. 1.12(a),(b) shows that there is a definite relationship involving the directions of the magnetic field, the current-flow, and the resultant motion. We can be grateful that nature has provided us with this very practical three-part interaction, but the two 'hand rules' are not arbitrarily interchangeable. Which one should be used?

The culprit here is the *direction* of current flow. Two viewpoints prevail in the technical literature. The modern viewpoint holds that current flow consists of negatively-charged electrons that leave the negative terminal of the DC source and complete the circuit by returning to the positive terminal. This is probably a physically-correct theory and a current so described is known as an *electron* current. However, it is a fact of life that much electrical phenomena has been dealt with for many years in terms of the older concept in which current flowed from the positive terminal of the DC source and returned to the negative terminal. Current described in this way is known as *conventional* current.

As electric motor technology is one of the older of the applied sciences, it is not surprising to encounter *both* viewpoints. As long as the author is consistent in dealing with the one or the other, there need be no violation of technical integrity. Problems arise when the reader isn't clearly and emphatically informed which viewpoint has been selected. The *assumption* is sometimes made that the electron current has made the conventional current obsolete. In other instances, the selected viewpoint appears in an obscure footnote all too easily glossed over by readers in pursuance of some particular data.

Note that (a) and (b) of Fig. 1.12 yield the same answers when two motor

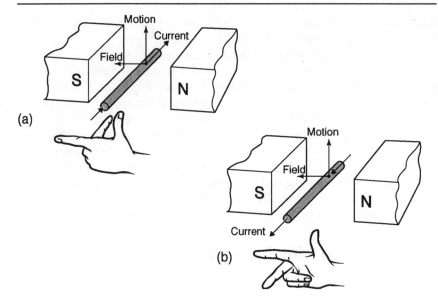

Fig. 1.12 *Resolving the confusion with the motor 'hand rules'. Some texts relate motor operating parameters with the left hand while others depict the use of the right hand.*
(a) The left-hand motor rule is valid for conventional *current-flow. (From 'plus' to 'minus'.)*
(b) The right-hand motor rule is valid for electron *current-flow. (From 'minus' to 'plus'.)*
In both cases, the thumb *indicates direction of motion, the* fore-finger *indicates direction of the magnetic field and the* middle-finger *indicates direction of current flow.*

parameters are known *with the proviso* that we must know which current flow viewpoint we are dealing with.

Force, current and flux – the orthogonal relationship

Besides indicating a third operational feature of motors when two are known or assumed, another characteristic is revealed by the hand rules. Note that the sketches always depict perpendicular displacements of the three digits. Indeed, maximum motor action occurs with the magnetic field and the current-carrying conductor perpendicular to one another. Moreover, the force thereby developed is perpendicular to *both*, the field and the current. This is known as the *orthogonal* relationship and is sometimes assumed in texts without further discussion. However, it should be known that this is not always achievable in electrical devices. It should be specifically pointed out that a heavy current-carrying conductor can be immersed in a

Electric motor generalities 19

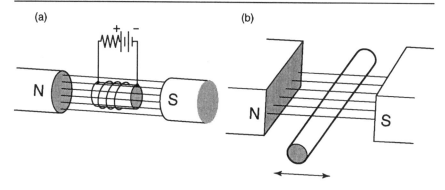

Fig. 1.13 *A magnetic field, conductor current, and motion do not always interact. These set-ups do not allow for an orthogonal relationship of flux, current and motion.*
(a) *The current-carrying conductor is situated with its longitudinal axis parallel to the magnetic field. No motor action takes place. (Also, if the DC source were removed and axial motion were applied to the conductor, no generator action would take place.)*
(b) *Here, again, the indicated motion applied to the conductor produces no generator action. From this, the inference can be drawn that motor action cannot stem from force acting parallel to the magnetic flux of the field-poles.*

strong magnetic field and *not* experience any displacement force or 'motor action' at all.

Imagine the situation suggested in Fig. 1.13(a). The current-carrying conductor is in the magnetic field, but its directional axis is aligned with that of the field. Not only is there no mutually-perpendicular relationship relating flux, current and force, there simply is *no* force developed on the conductor. (It is also true that motion *imparted* to the conductor along its longitudinal axis would induce no e.m.f. or voltage in the conductor, i.e. no generator action would take place.)

Further illumination of the idea being put across may be gleaned from the situation shown in Fig. 1.13(b). Here, again, no *generator* action is developed despite the motion imparted to a conductor immersed in a magnetic field. Inasmuch as generator and motor action always occur simultaneously (although one always predominates), we can infer the force developed on current-carrying conductors in motors is never *parallel* to the magnetic flux in which the conductors are immersed.

Aside from these extreme cases, one can have situations where the relationships of flux, current, and force deviate from the ideal orthogonal pattern. The practical aspect of such departure is that *less* force would be developed to produce the desired motor action. In DC motors, armature reaction can twist the direction of the field flux and thereby reduce the available torque.

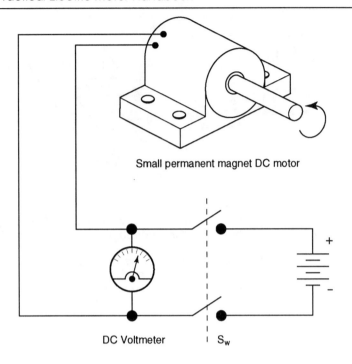

Fig. 1.14 *Simple demonstration of a basic operating feature of electric motors. Switch, S_w, is closed, placing the motor in operation. The voltmeter will indicate whatever voltage is applied to the motor. The switch is then opened whereupon it will be observed that the voltmeter indication remains close to its original value, then declines as the motor slows down. The interpretation is that the motor and the generator action occur* simultaneously. *Important, also, is that the generated voltage, i.e. the* counter-e.m.f., *is polarized so as to oppose the armature current delivered to the motor.*

Counter-e.m.f.

Most dynamos can be reciprocally used as either motors or generators. With electrical energy in and mechanical energy out, they function as motors; generator operation occurs by supplying, rather than loading, the shaft with mechanical torque, for then electrical energy is available at the former 'input' terminals. It is also true that motor *and* generator action are always present in an operational machine. These basic characteristics were demonstrated for a DC permanent magnet motor in the simple experiment depicted in Fig. 1.14.

A paradox arises if this demonstration is attempted with a squirrel-cage induction motor as shown in Fig. 1.15. We see no evidence of a counter-e.m.f. when the switch is opened – the AC voltmeter instantly returns to zero even though the unenergized motor continues to coast for a long time.

Electric motor generalities 21

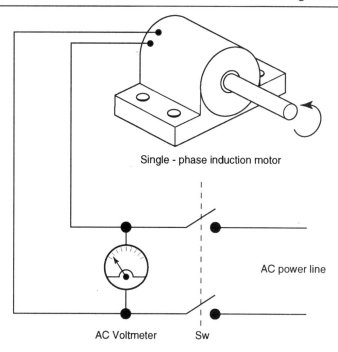

Fig. 1.15 *Set-up intended to show generator action in an induction motor. Unlike the experiment of Fig. 1.14, no generator action is revealed when the switch is opened. During the entire coasting period, the voltmeter indicates no counter-e.m.f. The explanation of this paradox is that both* the interacting magnetic fields in the motor are deactivated when the switch is opened. This *contrasts to the situation in Fig. 1.14 where only* one *field is deactivated. A corollary of the above experiment is that the induction generator is fail-safe in that it cannot deliver a short-current to a faulted power line.*

Intuitively, this is disturbing. We feel that *something* must prevent a normally-operating induction motor from consuming a near short-circuit line current. This can be inferred from the momentarily high line current that a polyphase induction motor draws when at standstill. ('pure' single-phase induction motors are not self-starting, but would draw a *sustained* high-current at standstill.) Where, then is evidence of the elusive counter-e.m.f. which we know must exist to account for the current limiting in a normally-operating induction motor?

This dilemma evaporates when we contemplate the full effect of opening the switch in Fig. 1.15. The armature of the motor is de-energized, but so is the interacting field from the squirrel-cage rotor. It is a matter of simple transformer action. Without interacting fields between the stationary and rotating members, a machine can function neither as a motor nor as a generator. The *practical* fact here is that if an induction motor is *supplied* with mechanical power and thereby made to rotate *faster* than the synchronous speed set by the number of poles and the line frequency, the motor becomes an *induction generator* delivering current *into* the AC power line. Here, then is our elusive counter-e.m.f. at work! Here, too, is a practical insight into the nature of induction motors. A nice thing about this phenomenon is that the frequency of such an induction generator is set by the AC power-line, but the greater the speed, the more current is pumped into the power line.

Electric motors in the quest for perpetual motion

A favourite perpetual motion system comprises an electric motor and a generator each supplying its partner's energy needs. Despite the inherent losses in each machine, the output shaft is supposed to rotate perpetually and/or supply useful mechanical energy while doing so. Generally, those who attempt to set the world afire with such a wonderful concept either subconsciously or intentionally complicate the idea by inserting various gearboxes, flywheels or other components between the motor and the generator. These, it is hoped, would somehow compensate for friction and electrical losses as well as for the laws of thermodynamics. To give credit where due, one must sometimes commend the ingenuity of these implementations. Sadly, however, the practical result of giving such a system an initial boost is a rapid deceleration to a standstill. Typical of such attempts to fool nature is the system shown in Fig. 1.16.

Some time ago a well-known popular science magazine appeared at the newsagents with an artist's rendition of a new 'electric' motor destined to completely revolutionize the nature of industrial society. This motor consumed no electrical energy at all; indeed it was not intended to be connected to the power line. The alleged inventor had observed the fact that motor action inevitably stemmed from the interaction of magnetic fields. The natural extrapolation drawn was 'who needs electricity? It can be done directly with magnets.' Fig. 1.17 exemplifies this fallacy.

Such motors, allegedly operated entirely by magnets remain popular pursuits by the fringe inventors' groups. This goal somehow appears more attainable than perpetual motion. Practical evidence of either continues to be absent.

Yet another not uncommon illusion should be shattered. Hobbyists have made simple DC motors/generators such as depicted in Fig. 1.18. It is easy enough to observe that stronger motor torque and/or speed can result from stronger magnets. As a generator, greater electrical output results from stronger magnets. Some have extrapolated this to the limit by interpreting the source of output energy to come from the *magnet*. In so-doing, they overlook the practical fact that the magnet does not get 'used up'. Actually the magnet is like a catalytic agent in chemistry – it promotes a reaction, but does not impart its energy to it.

Electric motor generalities 23

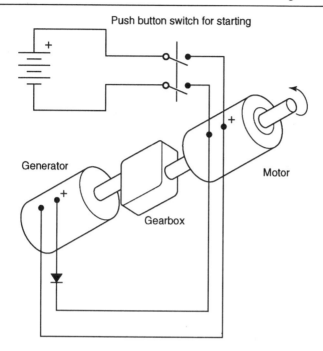

Fig. 1.16 *A typical electro-mechanical perpetual motion system. The truly perpetual aspect is the never-ending notion by enthusiastic, but misguided, inventors who continually advocate this solution to the world's energy crisis. Some even claim practicality to be 'just around the corner' – 'all that needs to be done is to eliminate friction'.*
The 'motor-feeding-a-generator-feeding-the-motor' scheme is presented in many guises and disguises. A favourite is to insert gearboxes for various alleged purposes. Mysteriously, a working model never seems to surface.

Idealized concept of energy conversion

The mind-set of those seeking to demonstrate perpetual motion arrogantly brushes aside nature's dictum that no machine, including electric motors, can make any kind of energy transformation with 100% efficiency. Why, indeed, do they behave as if they have a special dispensation to violate the inviolate laws of physics? A possible answer can be found in electric machinery texts which unambiguously state that the internal conversion of electromagnetic to mechanical energy in a motor is 100% efficient! Unambiguous as the language may be, problems arise only when it is taken out of context. For, these texts do not take issue with the *practical* fact that a 10 hp motor must always develop a *higher* internal horsepower, say 11 or 12 hp, and the input power must be higher yet.

In other words, the rated horsepower represents the power output of the motor that is *available* at the shaft to do useful work. The perfect conversion

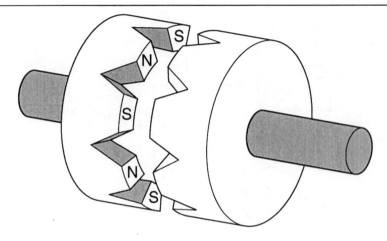

Fig. 1.17 General configuration of the long-sought 'magnetic motor'. Adament inventors claim to have produced working models but actual evidence has never been forthcoming. There are at least two difficulties with the concept of such a device. The continuous rotation would be prevented because of the tendency for the magnetic poles to lock-up. And if, as usually contended, the mechanical energy supplied by the shaft was derived from the stored magnetic energy, the magnets would be discharged long before any practical useful work was available.

from electromagnetic to mechanical energy cannot be experienced in practice for two basic reasons. First, there are inevitable electrical losses between the power line and the torque-producing process within the motor. Second, there are inevitable mechanical losses between the torque-producing process and the output shaft of the motor. It is therefore *always* true that the energy input to a motor equals the energy output plus the sum of all the losses. So, despite the claimed perfect-conversion *within* the motor, the efficiency must always be *less* than 100%, it is not possible to make practical use of the lossless transformation of energy *somewhere* within the motor. The energy-flow diagram of Fig. 1.19 illustrates these matters.

Interestingly, the mirror-image of this concept applies to electric generators. There, the mechanical losses are said to occur *ahead* of the energy conversion process. The electrical losses are then construed to take place between energy conversion and the electrical output of the machine.

Summarizing, we have a useful concept which provides convenient translation into practice. At the same time, one must not make the misinterpretation that it is possible to tap into the perfect energy conversion process. Suggestions such as this often appear in science and technology.

Electric motor generalities 25

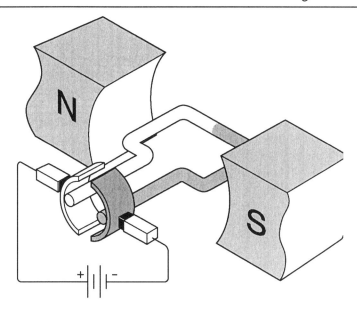

Fig. 1.18 *Simplified sketch of a basic DC electric motor. The essential items are the magnetic poles, the single-turn armature, commutator, brushes, and DC source. The exact configuration, bearings, shafts, etc. is not critical and will vary with the builders. It is, however, easy to misinterpret the operational features of the motor. A popular misconception attributes the mechanical energy developed to come from the* magnet. *(It is also easy to overlook that the armature of DC machines in operation always carries* alternating currents.*)*

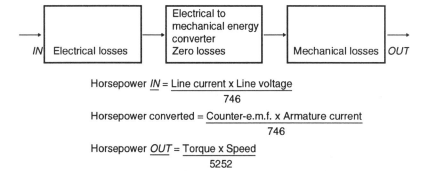

$$\text{Horsepower } \underline{IN} = \frac{\text{Line current} \times \text{Line voltage}}{746}$$

$$\text{Horsepower converted} = \frac{\text{Counter-e.m.f.} \times \text{Armature current}}{746}$$

$$\text{Horsepower } \underline{OUT} = \frac{\text{Torque} \times \text{Speed}}{5252}$$

Fig. 1.19 *Energy-flow diagram for electric motors. The conceptual representation of motor operation is very useful in making calculations. However, the energy converter must not be misinterpreted. The zero-loss feature simply tells us that energy cannot be created or destroyed, but only transformed in nature. There is no practical way to harness the 100% efficiency of the energy conversion process.*

Measurement pitfalls

Practical experience with motors teaches us that many useful objectives can be realized by dealing with reasonable approximations. Despite the liberal tolerances permissible in motor measurements and calculations, it is nonetheless profitable to at least be aware of the common sources of error.

The measurement of low-resistances, such as those of armatures comes to mind. Much, of course depends upon the nature of the instruments used. An easy pitfall here is the lead resistance and either the instrument or the operator must cancel the effect on the readout. Somewhat more subtle is the usual assumption that DC and AC resistance are the same. We are not alluding to inductive reactance, but rather to the so-called 'skin-effect' wherein the actual *resistivity* of a conductor becomes greater with AC because of the concentration of current flow in the outer periphery of the conductor. At 50/60 Hz, one does well to allot at least a 10% increase of the AC resistance over the DC resistance.

In the interest of working with realistic resistance measurements, it is also a good idea to make the measurements before *and* after the normal temperature rise has set in. Then one is in a better position to evaluate such performance parameters as starting torque, starting current, and speed regulation. Keep in mind too that the resistance of carbon and graphite brushes, unlike copper, *decreases* with increasing temperature.

Often overlooked is the deleterious effect on power line power factor when motor control is brought about by SCRs and Triacs, especially at low conduction-angles. Engineering texts usually deal with the power factor in sine-wave circuits containing inductive and capacitive reactance. Although not easy to grasp intuitively, it is a very practical fact of life that low power factor also results from voltage and current having *different* waveshapes. Interestingly, the *load* power factor in the SCR/Triac phase-control circuits is approximately unity for any conduction angle because the load current and the load voltage, although non-sinusoidal, are nearly the *same*. Insight into these matters are provided by Fig. 1.20 and Table 1.1.

The electric vehicle

The choice of motor best suited for electric vehicle propulsion has been, and remains a controversial topic. Before solid-state power equipment became available, the logical selection was the DC series motor. Aside from the limited options from which one could then choose, the DC series motor had the compelling feature of high starting and low-speed torque. However, speed control by rheostat or by tapped switches was both inefficient and impractical. A quantum leap in electric vehicle technology occurred when it became feasible to use chopper duty-cycle control or pulse-width modulation to control speed and/or torque. It was also found that very good results

Electric motor generalities 27

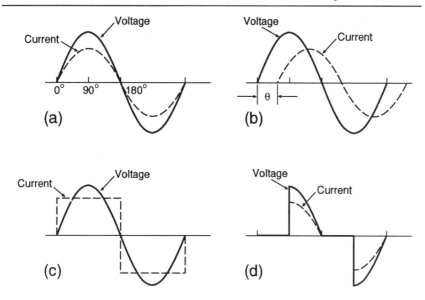

Fig. 1.20 Deceptive nature of non-sinusoidal waveshapes. The traditional concept of power factor involved only sine waves. Power factor obeys different rules when non-sinusoidal waves are involved.
(a) and (b) Power factor as governed by inductance and capacitance in sine-wave circuits. Unity power factor exists in (a). In (b) the current lags the voltage by angle θ and the power factor is cos θ.
(c) and (d) Non-sinusoidal waves with no inductance or capacitance. Power factor in (c) is less than unity despite 'in-phase' condition. Power factor in (d) is unity because of identical waveshapes.

were obtained by employing the permanent magnet DC motor in such a system.

The desire to dispense with the maintenance problems of commutator–brush machines ushered in the development of various AC motor drives, together with the DC brushless motor. The heart of such systems is the solid-state inverter and dedicated IC control modules. Such motors tend to be lighter and more efficient than the classic DC types and, moreover, they make it easy to build the motor actually into the wheels of the car. There is, overall, a certain elegance to such a system that appeals to the design engineer. Nonetheless, when *all* vehicular parameters are balanced, valid arguments still persist for *both* DC and AC motors. The difference in performance, range, maintenance and cost is not yet overwhelming in favour of either drive technique.

It turns out that the *main* bottleneck in public acceptance of the electric car is the limited range between charges. This translates into the energy storage capability of the battery. By now we recognize that the tauted

1	2	3	4	5	6	7	8
LINE VOLTAGE WAVEFORM (SINE WAVES)	CURRENT CONDUCTION ANGLE IN DEGREES	LINE AND LOAD CURRENT WAVEFORM. (LOAD VOLTAGE WAVEFORM.)	EFFECTIVE CURRENT (RMS VALUE)	EFFECTIVE VOLT-AMPERES	ACTUAL POWER IN LOAD RESISTANCE	POWER FACTOR AT AC LINE.	POWER FACTOR AT LOAD
∿ E_m	180	∿ I_m	0.707 Im	0.500 EmIm	0.500 EmIm	1.000	1.000
∿	150	∿	0.697 Im	0.493 EmIm	0.486 EmIm	0.985	1.000
∿	120	∿	0.634 Im	0.448 EmIm	0.402 EmIm	0.897	1.000
∿	90	∿	0.500 Im	0.354 EmIm	0.250 EmIm	0.707	1.000
∿	60	∿	0.313 Im	0.221 EmIm	0.098 EmIm	0.443	1.000
∿	30	∿	0.120 Im	0.085 EmIm	0.014 EmIm	0.170	1.000

Table 1.1 *Circuit values when a load is controlled by a triac or anti-parallel SCR. Non-sinusoidal waveshapes are produced by popular electronic motor-control circuits predicated on full-wave phase variation. The line power factor consequences depicted in column 7 often evoke surprise. These are lagging power-factors. On the other hand, the load power factors listed in column 8 remain immune to conduction angle. This situation is exemplified by the waveshapes shown in (d) of Fig. 1.20. (Keep in mind that rms-responding instruments are needed for such circuits.) SCR half-wave control produces even poorer AC line power factors.*

performance of various exotic batteries, although confirmed in laboratory tests, inevitably prove either uneconomical or impractical, and usually both, in the actual driving environment. Pending some unexpected breakthrough, it appears that electric propulsion will first appear in *hybrid* systems in conjunction with small internal combustion engines, or will benefit from practical and economic improvements of fuel cells. Here, hydrogen would be the ideal fuel, but various hydrocarbon liquids or gases may serve as acceptable compromises.

Despite the on-going controversies regarding electric motor types and battery formulations, the advantages of the hybrid format over the 'pure' electric vehicle has never been in doubt. Indeed, it appears to be even more compelling as exotic batteries and avant-garde energy storage devices continually fail to meet the harsh requirements of the practical world.

It appears that the basic problem involved in developing an economically-successful hybrid vehicle is to find the optimum balance between the ratings and deployment of the combustion and the electric power sources. It does seem clear that a relatively small contribution from the internal-combustion engine goes a long way in 'practicalizing' the hybrid vehicle. This pertains to extended range, enhanced convenience in battery charging and in insurance against being stranded when far from home. And, regardless of the configuration of the successful hybrid vehicle, its contribution to atmospheric pollution is bound to be small relative to that of present petrol- and diesel-powered automobiles.

So much focus has been directed upon the technical aspects of alternatively-powered vehicles that other very important things have been brushed aside. There are political and psychological forces to consider. Consumers will tend to compare performance parameters with those of conventional automobiles and long-enduring industrial activities cannot be rendered obsolete overnight. Everywhere, there will be vested interests to contend with; those seeking to profit from the new technology would probably be wiser to represent it as filling a niche rather than replacing an entrenched industry.

The two major types of hybrid schemes are shown in Fig. 1.21. In the series type, the wheels are driven by the electric motor. The engine–generator set can be used not only to charge the battery but, bypassing the battery it can power the electric motor. In the parallel hybrid, the electric motor, the engine, or *both* can drive the wheels. With the electric motor acting as a generator, the battery can be charged.

Inasmuch as the design of electric vehicles brings together the techniques of solid-state electronics, electric power engineering, computer logic and the extensive experience of the automotive industry, it should not be too surprising to encounter some rather novel approaches. Consider, as an example, the implementation of regenerative braking. The basic idea here is to recover some of the energy which would otherwise be lost as heat when

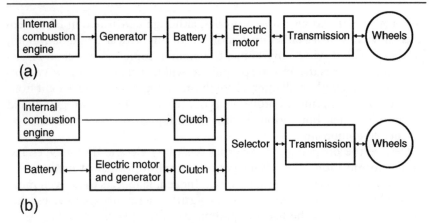

Fig. 1.21 *The two major formats for hybrid vehicles.*
(a) The series hybrid: Only the electric motor drives the wheels. The engine–generator combination can be used, however, to bypass the battery and directly power the electric motor.
(b) The parallel hybrid: The engine, the electric motor, or both may be used to drive the wheels. The electric motor, acting as a generator in conjunction with the engine, can charge the battery. With hybrid technology, the type of electric motor used is less important than in pure electric vehicles.

braking or decelerating. By converting such frictional and kinetic energy into electricity, the battery can be charged, thus extending range and also reducing the wear on the mechanical brakes.

Traditionally, regenerative braking has usually been accomplished in electric cars by taking advantage of the fact that DC motors conveniently act as DC generators when mechanical torque is imparted to, rather than extracted from, their output shafts. However, in some modern electric vehicles propulsion results from AC induction motors built into the wheels. Here the mechanical engineers have evidenced their contribution in the interest of reducing weight and manufacturing costs and increasing efficiency. The question arises as to how regenerative braking can be realized from such a propulsion system.

Interestingly, electric power engineers have long known that the induction motor becomes an alternator when its shaft is driven above the synchronous speed of its rotating electromagnetic field. This is not always mentioned in handbooks because the phenomenon had not been extensively exploited in utility systems. Actually, it is a nice sort of behaviour because it occurs automatically and the reversed flow of energy obligingly takes place at the same frequency that is impressed on the motor and without the need to meet any phase demands.

Next, the electronics engineers developed variable-frequency DC to AC inverters that allowed energy transfer in both directions – to and from the battery. Finally, the computer specialist left his hallmark – a myriad of system logic and automated control functions. With such viability, have we been focusing on the wrong horizon?

Things to keep in mind about motors

It has been the author's observation that otherwise competent engineers sometimes engender needless trouble by treating electric motors as 'just another passive device'. It is always important to keep in mind that motors are *dynamic* devices with sometimes wildly-varying characteristics, which are functions of time, circumstance, and perhaps 'happenstance'. (One cannot predict, for example, the state of residual magnetism in certain machines. The condition of the brushes and commutator may cause excessive radio-frequency interference before any adverse change in motor performance is detected.)

In most motors of a substantial fraction of a horsepower, and certainly with those rated above several horsepower, the current inrush at start-up must be dealt with. Some current-limiting technique is called for to protect the motor and to prevent disturbances to other loads feeding from the power line. Exactly where one draws the line depends upon the feeder-line characteristics and somewhat on the nature of the load being handled. For example, a high-inertia load will prolong the motor acceleration time, thereby increasing the energy content of the inrush transient.

Nature endows DC motors with an internal feedback mechanism which expresses itself as the tendency for the counter-e.m.f. to *almost* equal the no-load applied voltage. Then, the *difference* between the two voltages becomes just enough to permit the required current and therefore, torque for the load being imposed. AC motors operate in a similar fashion, except that the torque-producing current is also regulated by *power factor*. The common denominator between DC and AC motors is that they both automatically allow more input power in order to accommodate increased load. Both seek and find a steady operating equilibrium.

Keep in mind, too, that an increase in the developed torque tends to cause higher speed. It is not true, however, that torque is dependent upon speed itself. Torque and speed change may derive from common cause, but the relationship remains unilateral as described. Finally, keep in mind that during operation, motors are *simultaneously* acting as generators. This awareness, alone, leads to meaningful insights and practical control techniques. So much for generalities; let's now take a closer look at these motors.

2 Practical aspects of DC motors

Background of DC motors

DC motors, because of their brush–commutator systems, are more complex dynamos than those which operate from AC. An exception might be the Faraday disc which provided continuous rotation when powered from a DC source. Several practical difficulties conspired to prevent the Faraday disc from becoming an industrial workhorse, however. Chief among these was the inordinantly high current and impractically low voltage combination needed. Pre-dating the Faraday disc, a commutator–brush motor was reputedly demonstrated by an American blacksmith in the mid-1830s. Largely because of the inadequate battery technology of the era, the invention attracted attention only as a curiosity.*

Once DC motor technology got under way, it matured rapidly and was conveniently available to meet such heavy demands as those imposed by the automobile industry, garment manufacturing and, when deployed 'backwards', for generating power for street lighting and general electrification purposes. In relatively recent times electric motors, in general, were often viewed as mundane if not boring technologies with little possibility of much further evolution. This attitude is now *passé*.

Today, DC motors once again command widespread interest as exceptionally useful devices. This dramatic change has been brought about by the advent of solid-state rectifiers, new and exotic magnetic materials, electronic control techniques, electric vehicles, computers, etc. Last, but not least, the DC series motor became the universal motor, operating from both DC and AC with very minor modification.

---

*The blacksmith, Thomas Davenport, patented his 'electromagnetic engine' in 1837. Despite capable collaborators and a financial backer, commercial success was never realized. Davenport died pennyless in 1851. His invention was too far ahead of the times.

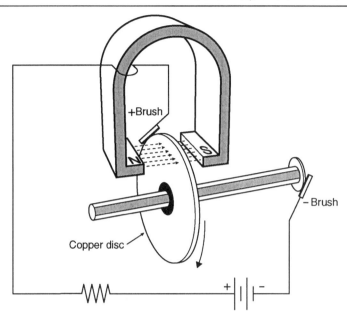

Fig. 2.1 *Michael Faraday's homopolar motor. Known also as an acyclic machine or a Faraday's disc, it is reputed to be the world's first type of electric dynamo. It functions equally well as a motor or a generator and stands unique in that no alternating current is involved in its operation, so no commutator is needed. It remains a very practical machine where high current and low voltage is either available or required.*

The homopolar motor

One is hardly likely to encounter the homopolar motor/generator around the house as an appliance motor. Some people have classified it as an archaic machine because of its status as one of the very earliest electric dynamos. Actually, its apparent rarity stems from its specialized applications. Moreover, because of modern materials, solid-state techniques and superconductivity considerable impetus has been given to improved versions of this relatively simple motor. Most important, however, are the insights to be derived from an understanding of its operation, see Fig. 2.1.

As with most electric motors, this dynamo is either a motor or a generator depending upon whether the input power is electrical, or mechanical. In either case as with nearly all other machines motor and generator action also occur simultaneously. A significant *difference* from other motors is that the disc-type armature carries only DC; in other motors, whether DC or AC, the armature has alternating current in it and for this reason, no commutator is needed. Contact to the shaft and disc is made via slip rings, brushes or sometimes liquids.

This motor can be constructed in large sizes, such that tens-of-thousands of horsepower are available for applications such as ship propulsion. As a generator, thousands of amperes of DC current can be produced, but only at a few volts or so. Multi-disc design can only partially overcome this negative feature. Low voltage operation occurs because the disc comprises the equivalent of just one-half turn of armature winding. Armatures of other motors usually have multi-turn windings.

'Homopolar' does not imply a single magnetic pole but rather it signifies that there is just one interaction of the active segment of the armature per revolution. In all other motors, there are two or more. Interestingly, the magnet-hydrodynamic generator works on the very same principle, but uses a hot jet of plasma (ionized gas) in place of the copper disc.

The AC involvement in DC motors

A simple experiment on the nature of motor action should be performed by the hobbyist and professional alike, for some vital insights are hard to come by through either mathematical analysis or intellectual reasoning. In the set-up of Fig. 2.2 a DC shunt motor is being operated in nearly normal fashion; the one minor deviation from everyday practice involves the presence of the inductor L in the shunt field circuit. If, however, this inductor consists of a number of turns of heavy gauge wire wound on an iron or ferrite core there need be no disturbance of the ordinary motor operation.

The inductor has been inserted to enable oscilloscopic examination of AC voltage induced in the shunt-field winding. (Without the inductor, such an induced AC voltage is likely to be heavily-loaded by the shunt field DC source.) 'Commonsense' would support the existence of such an induced AC voltage for the simple reason that the armature of all DC machines actually carries alternating current. Certainly, our knee-jerk intuition would suggest there is plenty of iron available to form good magnetically-coupled circuits. In a way, this anticipated induced voltage appears somewhat similar to the counter-e.m.f. of the motor. Better still, its existence would be attributed to simple *transformer* action.

As the reader will no doubt have by now suspected a paradox, we can now acknowledge that almost no such induced voltage appears in the shunt-field winding. A unilateral situation exists in which the armature current in the presence of the field flux produces motor action, but practically no induced AC voltage appears in the field winding from the armature current. Despite physical proximity and the abundance of iron in both armature and field poles, the *spatial orientation* of the two fields is 90° from what it would have to be to produce maximum electromagnetic coupling. As it is, there is zero (or nearly so) coupling and there is almost no transformer action. It is analogous to two solenoids in which no mutual

Practical aspects of DC motors 35

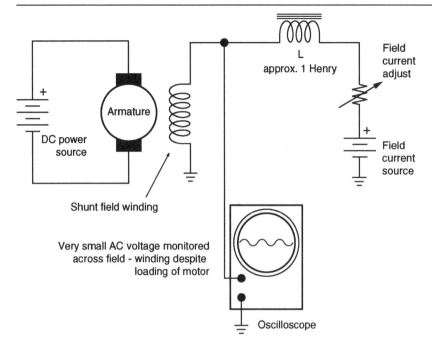

Fig. 2.2 *Simple set-up to disprove transformer relationship between armature and field winding. Despite the fact that the armature carries alternating current and is physically close to the field poles, it does* not *act as a transformer primary with the field winding as secondary. Thus, in the above set-up, very little AC voltage is monitored across the field winding. Moreover, loading the motor to increase armature current causes very little increase in the small AC voltage 'scoped' in the field winding. (The inductor, L, is inserted to provide isolation so that if there were a transformer-induced voltage in the field winding, it would not be loaded down by the DC field-current source.)*

coupling exists when one solenoid is perpendicular to the axis of the other. This is a very fortunate situation for motors. Let's see why.

First, if the speculated transformer coupling between the armature conductors and the field winding existed, the nominal DC field current would be AC modulated and it would be most difficult to postulate the effect on motor performance. Suffice to say that it doesn't appear to be favourable. Moreover, it is likely that a permanent magnet DC motor wouldn't be practical – the magnet would be subjected to the demagnetizing effect of an AC field.

Secondly, we would find the operating principles of a number of unique machines upset. This would include repulsion motors, amplidynes, the rotorol and regulex and the Rosenberg generator. The space–quadrature relationship between the magnetic flux of the current-carrying armature conductors and the main field flux is one of the facts of life of many types of motors.

It is now necessary for the author to defend himself from attacks from purists who will certainly point out that despite their quadrature relationship, the two magnetic fields do more to one another than just develop motor action. Admittedly, we have dealt with an 'ivory-tower' motor; real-life motors tend to suffer from both diminituon and distortion of the main field air-gap flux. This is known as *armature reaction* – the influence of the armature field on the field produced by the pole windings. Because of armature reaction, motor performance, especially commutation, can be severely degraded. *Practically*, however, the effects of armature reaction are largely designed out, cancelled or otherwise rendered negligent.

This, indeed is what presents a challenge in motor design. Many techniques are used such as compensating windings, interpoles and semi-empirical modifications of the structure and shape of the field poles. Shifting the brush axis from geometrical 'neutral' may be necessary. In practice, therefore, motor behaviour can closely approximate that of an ideal machine in which armature reaction never asserted itself in the first place. The experiment of Fig. 2.2 bears this out, for even when the motor is heavily loaded, there is little evidence of transformer coupling from the armature to the field winding.

A practical view of armature reaction

The experiment described for the set-up of Fig. 2.2 and the ideas associated with it bring us to a very practical conception of the inner workings of the electric motor. Whereas engineering texts generally treat basic motor action, such as that depicted in Fig. 2.2 and armature reaction separately, our practical purposes are better served by viewing this phenomena as just *one* cause and effect. The cause is the current in the armature conductors which thereby produce its own magnetic field. The effect is the distortion of the main air-gap field due to interaction between the magnetic fields of the field winding and the armature conductors. The key word here is *distortion*.

As a word, distortion usually has negative connotations. On the one hand, however, our much-desired motor action stems from the type of distortion shown in Fig. 2.2. This, however, is not an overall accurate depiction of the field distortion tending to be produced in practical machines. It is more on the order of the twisted flux lines shown in Fig. 2.3. This is said to emanate from *armature reaction*, as if a separate cause now produces a separate effect. This is fine for the purposes of mathematical analysis and as a teaching mechanism. Our practical insight is best served, however, by considering the air-gap field distortion as the immediate cause of *both* motor action *and* the deleterious consequences of armature reaction.

The nice thing about this viewpoint is its simplicity. It turns out to be

Practical aspects of DC motors 37

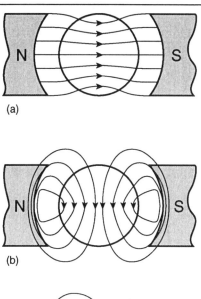

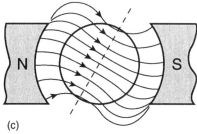

Fig. 2.3 *Armature reaction–flux distortion is bad for motor and generator action. In addition to the desirable field flux distortion the general twisting of the magnetic lines of force tends to occur.*
(a) Simplified view of the magnetic field in a DC shunt machine with field winding energized, but with no armature current.
(b) Simplified view of the magnetic field produced by armature current with zero field current.
(c) Distortion of the composite field due to interaction of the fields shown in (a) and (b). Note: Practical machines incorporate techniques to restore flux-pattern (c) very nearly to that of (a).

simple because practical motors incorporate means to greatly *minimize* the effects of the field distortion attributed to armature reaction. (If this were not so, the set-up of Fig. 2.2 would have revealed a large AC voltage induced in the field winding.) Interestingly, a tiny bit of the armature reaction type of distortion may weaken the air-gap field a bit when the motor is heavily loaded and thereby *improve* the speed regulation. However, this technique whether inadvertent or otherwise can also degrade commutation or can promote instability. (The motor 'thinks' it is differentially-compounded.)

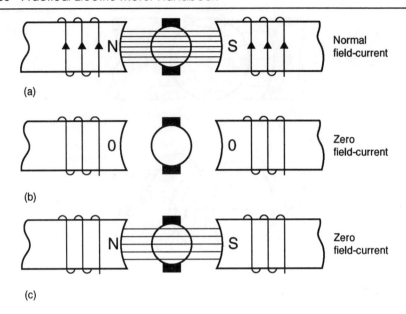

Fig. 2.4 *Simplified diagrams of the polar field-flux in a DC shunt motor.*
(a) Normal field current in shunt field windings. Note density of magnetic field in air gap.
(b) Erroneous assumption: The magnetic flux from the poles does not vanish with the cessation of field current in the shunt windings.
(c) Actual situation: Because of residual magnetism, a weak field persists even though there is no field current. This is the key to the runaway characteristic of DC shunt field motors.

The role of residual magnetism

Despite the logic, equations, and examples advanced to clarify the paradox of the speed behaviour of shunt motors, the student first encountering the topic may remain unconvinced. It would be natural enough for him to reason that a DC shunt motor deprived of field current should come to a halt rather than race. In other words, no field, no motor action. Unfortunately, this assumption is fostered by the wording found in some text books which convey the notion that zero field current and zero magnetic field flux in the air-gap occur together. If this were so, the student's intuition would be valid and no torque would be developed in the armature once the shunt field current was interrupted.

What these texts either fail to mention or emphasize sufficiently is that there is *residual magnetism* left in the pole pieces. This being so, zero field current does *not* interrupt the magnetic field, but merely *weakens* it. See Fig. 2.4. The resultant reduction in counter-e.m.f. then allows a higher armature current, developing increased torque which accelerates the motor to higher speed. At some dangerously-high speed, the counter-e.m.f. can again become high enough to slow the acceleration and approach an equilibrium between torque and load. Emphatically, the runaway response is not due to a *lack* of magnetic field, but to a *weaker* than normal field.

A practical demonstration of the residual polar field in DC machines is to operate the shunt motor as a self-excited generator. If everything is in order, rotation supplied to the shaft will cause the generated voltage to build up from a low value to a sustained output level. The initial low value stems from the cutting of residual magnetic flux by the armature conductors. If, because of vibration, temperature rise or some operating condition, the normal residual magnetism is not present in the poles, there will be no build-up process and no output voltage will be generated. For the sake of experiment, one can virtually wipe out the residual magnetism by applying an AC voltage to the field winding of the stationary machine and gradually reducing this voltage to zero.

The DC shunt motor

In the technical literature, we find common reference to the DC shunt motor as a constant speed machine. This it is not in the absolute sense – only the AC synchronous motor maintains an average constant speed in the face of load and voltage variations. Relatively speaking, however, the DC shunt motor (and the AC squirrel-cage induction motor) display exceedingly good speed regulation. In contrast, other motors, such as the DC/AC series motor and some DC compound motors show rather drastic changes in speed as their operating conditions change. The point is that for many *practical* applications, the DC shunt motor serves its purpose almost as well as if its speed were inviolate. Performance curves are shown in Fig. 2.5.

It is appropriate to reiterate the essence of our previous discussion on residual magnetism. Some students strongly feel that a DC shunt motor should come to a standstill as a consequence of opening the field current circuit. The rationale is, of course, that motor action must disappear in the absence of polar magnetic flux in the air-gap. As has been pointed out, however, residual magnetism remains in the pole-pieces despite zero current in the field winding. In practice, therefore, the motor experiences a *weakened* field, *not* an interrupted field.

Hammering home this fact of life, may not make it easier to grasp the important fact that a weakened magnetic field brings about *higher* rather than lower speed. The latter idea is not altogether faulty, for, as we shall see, there are DC shunt motors that behave in inverse fashion to the more conventional machines. It has to do with the presence or absence of magnetically-permeable material (such as iron) in the overall magnetic circuit. With conventional motors the iron structure makes the motor's counter-e.m.f. the dominant parameter in speed determination. A *little* weakening of the

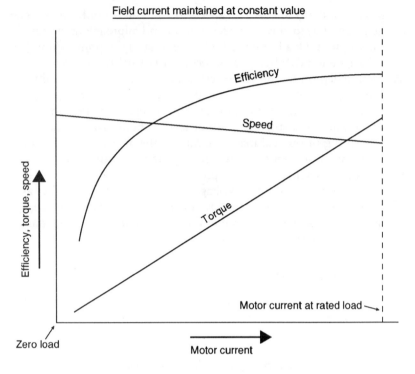

Fig. 2.5 *General characteristics of the DC shunt motor. Motor current is adjusted by varying the mechanical loading. Salient features of the DC shunt motor are: (a) Good speed regulation – this motor is referred to as a 'constant-speed machine', (b) a small change in field current produces a large change in torque, (c) will not run on AC, (d) reversal by transposing either armature or field connections, (e) runaway tendency at zero field current.*

polar field flux *greatly* reduces the counter-e.m.f. In turn this allows a large inrush of armature current torqueing the motor to *higher* speed. The motor tends to re-establish equilibrium, speeding up to raise its counter-e.m.f. thus slowing down the increase of armature current. We will find ourselves harking back to this basic theme in further discussions of DC motors.

A salient feature of the DC shunt motor is its reversibility. The direction of rotation may be reversed by changing the direction of current flow in *either* the shunt field winding or in the armature, but *not* in both. Moreover, such rotational reversal can be accomplished by 'plugging', that is by reversing one of the currents while the motor is in normal operation. In this way, it becomes unnecessary to wait while the motor coasts down to a standstill.

At first thought, it would seem best to reverse the shunt field current because it is very much smaller than the armature current. This may indeed be satisfactory, but several factors merit some consideration. The field winding is highly inductive; at the very least, trouble may be encountered with destructive arcing when the circuit is switched. In large machines, the high voltage thereby developed can be hazardous to both life and insulation. There is also the possibility of problems with interpole and compensating windings in some machines if strict attention isn't paid to current direction in such windings.

Transposing the connections to the armature tends to be a more straightforward reversal technique and except in very large motors, the momentary switching of the armature current should pose no difficulties. It is not uncommon to find that a motor performs more favourably in one direction of rotation than the other. The disparity is usually evidenced by excessive brush-sparking in the less-favoured direction. Such commutation trouble may be remedied by shifting the brush axis, where this is feasible. Trouble of this nature is most likely to be encountered when striving for a wide range of speed adjustment – it is not easy to exceed a four-to-one speed range via field current adjustment.

A common cause of unequal performance in the two directions of rotation is misbehaviour of the load. Its torque demand for the two directions of rotation may not be the same. This, of course, is a mechanical not an electrical problem. If the load has symmetrical characteristics the speed regulation of the motor should be nearly the same in both rotational directions.

Another argument against switching field current to reverse the direction of rotation in the DC shunt motor is the danger of runaway if the switch becomes defective and deprives the field winding of current. Of course, the larger the machine, the more one must be concerned with the possibility of such racing and the accompanying heavy line current. At the other extreme, small sub-fractional HP motors often incorporate enough bearing friction and windage to impose a practical limit to attainable speed.

If excessive weakening of the shunt field can ultimately result in instability and in racing, it would be only natural to ponder the effect of over-excitation of the field. In other words, how much can we slow the motor down by increasing the field current? Here nature takes a hand by preventing the progressive increase of the magnetic field strength. The barrier is magnetic saturation of the pole pieces and associated magnetic circuit. We simply don't have the option of indefinitely producing greater flux density in the air-gap. The primary result of exceeding rated field current is to produce temperature rise from the I^2R dissipation in the field winding. Actually, motor designers have sought to specify a control range which uses the linear region of field current versus magnetic field strength.

It fortunately happens that violent destruction of large DC shunt motors doesn't occur every time there is loss of field current. On the one hand, the accompanying rise of armature current opens the circuit breakers between the motor and the DC source. On the other hand, the resistance of the DC line itself may be sufficient to limit the motor's acceleration and ultimately limit its top speed. This is particularly true were the feeder conductors of the DC line heat up appreciably, thereby further increasing their protective resistance.

Keep in mind that the racing or runaway tendency exists with or without load – it just takes longer when driving a load. Also, before racing sets in, excessive sparking of the brushes may serve as a warning not to further weaken the shunt field.

The DC permanent magnet motor

The reputation of permanent magnet (PM) DC motors has been slowly catching up with their present capabilities. Earlier versions were relatively large and heavy and did not yield sterling performance. For example, their starting torque was not impressive and the mere fact that they eliminated the shunt field winding and its usually negligible power dissipation was not a compelling feature. Moreover, there were reliability problems pertaining to vulnerability to irreversible demagnetization of the PM pole structure.

More recently, PM motors have benefited greatly from advanced technology ceramic and alloy magnetic materials. It is simply wrong thinking to conceive of the modern PM motor as a 'lazy-engineer's' substitute for a shunt motor. On a horsepower output basis the PM motor can now be appreciably smaller and lighter than an equivalently-rated shunt motor. The starting torque can be several times greater than a shunt motor with otherwise-similar ratings. Another nice feature of the newer PM motors is that their speed–torque relationship is very linear over a wide range and is therefore easy to predict.

Neither designers nor users should abide by their previously-justified feelings about demagnetization. One could devise a scheme of abuse to deliberately impair the magnet's strength, however, enough practical experience has been accumulated to show that this is no longer a danger even with heavy-duty service.

The alluded characteristics of the DC PM motor are depicted graphically in Fig. 2.6. It is interesting to observe from Fig. 2.6(b) that, for a wide range of applied armature voltages, the speed–torque characteristics remain linear all the way down to a standstill as more torque is delivered. Such performance would have made designers of shunt motors very happy indeed.

Permanent magnet DC motors tend to exhibit better (lower) speed regulation than do shunt motors. In this regard, they compare favourably with AC induction motors. Their operating efficiency, too, tends to exceed that of shunt motors. Not only is the PM motor amenable to armature–voltage control of its speed, but usually its good performance is also continuous when its direction of rotation is reversed by reversal of the polarity of the applied armature voltage. It has also been observed that when the PM motor

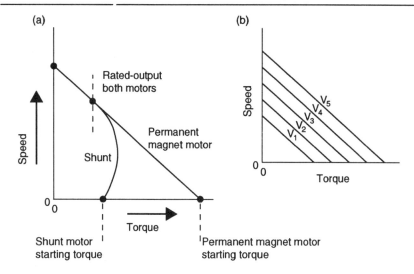

Fig. 2.6 *The unique torque–speed characteristics of the permanent magnet motor.*
(a) General comparison with a similarly-rated DC shunt motor. The salient feature of the PM motor is its relatively-high starting torque compared to its 'cousin', the DC shunt motor.
(b) Performance of the PM motor with different armature voltages. These voltages are progressively-higher from V_1 to V_5. Note the linearity. This behaviour is desirable for servo motor and instrumental applications.

is controlled by an audio-frequency pulse-width-modulation wave, operation is quieter than it would be in a similarly-controlled shunt motor. There, the need to use higher operating frequencies for the sake of noise-reduction can be electrically undesirable.

An important practical benefit of making use of a permanent magnet field structure is immunity to runaway. As is well known, if for any reason the DC shunt motor is deprived of field current the motor goes into a runaway mode. Even if it is fully loaded, the racing, together with abnormally-high armature current can pose a hazardous situation.

Motor designers attribute the excellent operating features of the PM motor to improved magnetic materials, which in turn has led to more favourable physical arrangements of the motor elements. In a manner of speaking, the PM motor can be thought of as a more-idealized implementation of the shunt motor. This enables one to appreciate why commutation and instability problems are relatively benign. It is known that such shortcomings in the shunt motor stem from the effects of armature reaction. Accordingly, it is safe to infer that the stiff air-gap field within the PM motor is not greatly affected by armature reaction which has long been the plague of motor designers.

Paradoxes in DC motor theory

It is easy enough to memorize precautions. The DC shunt motor can accelerate to an abnormally high speed and draw a dangerously high current if the shunt field winding is deprived of current and the series motor tends to race if deprived of its load. In order, however, to acquire a practical feeling for the causes and nature of such behaviour, we must carefully sort out the relationships, some of which are more subtle than obvious. Neither a quick glance at an equation, nor falling back on common sense is likely to provide the relevant insights. Certain apparent paradoxes and contradictions will be found most amenable to solution via a balanced blend of logic and intuition. More specifically, we will investigate the divergent performance of three types of DC shunt motors. In the conventional DC shunt motor, speed goes down in response to increased field current. Yet, in two other situations involving shunt motors, an *increase* in speed is brought about by more field current. Surprisingly, the same equation predicts this contradictory behaviour.

If we were motor designers, the exact speed of any of these DC machines could be calculated from an equation involving the following information:

(a) The voltage impressed upon the armature.
(b) The resistance of the armature.
(c) The voltage drop across the brushes.
(d) The number of poles.
(e) The number of active armature conductors.
(f) The strength of the net magnetic field in the air-gap.

Interestingly, our rather complicated equation would be applicable to all types of DC motors. Inasmuch as we will be working with already-designed machines, our interest will focus on trends and approximations. Despite the mathematical simplicity of this practical approach, some deep thought will be required to grasp the basic cause and effect relationships. A perusal of engineering texts will show that the resolution of the shunt motor paradox is not obvious from inspection. Yet, once a working insight is gained, it should pave the way for more successful implementation of DC motors in practical systems.

Consider the following relationship for conventional DC motors:

$$S = \frac{V_a - I_a R_a}{k\phi}$$

where S represents the speed,
R_a is the resistance of the armature,
I_a is the armature current,
ϕ is the strength of the magnetic field flux between the poles and the armature.

Practical aspects of DC motors 45

k takes into account the factors (d) and (c) which are fixed in a given machine. For our practical purposes, simplification can be affected by simply assuming that $k=1$.

Note also, that factor (c) can be assumed to be negligible.
This brings us to

$$S = \frac{V_a - I_a R_a}{\phi}$$

We should note that $V_a - I_a R_a$ is the counter-e.m.f. E_c, of the motor.
Accordingly, we can also deal with

$$S = \frac{E_c}{\phi}$$

It should be remembered that we are seeking trends of responses, *not* absolute values.

In words, the speed change in conventional DC motors is directly proportional to the counter-e.m.f., but inversely proportional to the strength of the field. This seems straightforward enough, however, when numbers are into the speed equation, one must always if a change in the denominator *also* engenders a change in the numerator (or vice versa).

These matters will be exemplified in the ensuing example. It will be seen that merely grinding the crank of an equation does not suffice; some practical aspects of motor behaviour must also be given due consideration. With this in mind, a practical question could well be, 'what happens to the speed of an operating DC shunt motor (of the conventional kind) when the DC field current is cut off?'

Superficially the answer appears easily forthcoming because a quick glance at the speed equation shows that S, the speed, increases if we reduce the field strength, ϕ. This, however, is not quite the correct answer; our practical experience reveals that stating that the speed merely increases is an understatement of fact. As pointed out, the reduction of field current to zero can result in *acceleration* of speed to the point of self-destruction. Also, it is an observable fact that armature current, I_a, *also* can become destructively large while the motor is thus accelerating. Our analysis now becomes a bit more involved.

What we must do is realize that a DC motor is simultaneously a DC generator; that is, *both* operational-modes occur together. The very existence of the counter-e.m.f. E_c, bears this out. It is not enough to ask what happens to motor action when the field strength, ϕ is reduced. One must *also* enquire what happens to generator action as revealed by a change in the counter-e.m.f., E_c and one must then ask what effect a change in E_c will have on *other* parameters, such as the armature current, I_a. Actually, the philosophy of this analysis will be seen to be quite straightforward and will resolve paradoxes that have been an obstacle to motor users.

Now, when we reduce ϕ, there must also be a proportionate reduction in the generated counter-e.m.f. This is a common-sense statement which stems from the nature of electromagnetic generation of voltage – such a voltage is proportional to the field strength *and* to the rate at which the magnetic flux of the field is cut. From here, cause and effect relationships may tend to be more subtle than obvious. Let's deal with the situation on a step-by-step basis.

Let us suppose a DC shunt-motor is *running* and its shaft is carrying, say about half of its rated load. The current in the field winding is interrupted and remains off. What happens?

(a) The counter-e.m.f. E_c, *instantly* drops in response to the reduced magnetic field flux. Note that the residual magnetism in the poles prevents the field from falling all the way to zero, even though there is now zero current in the shunt field winding.

(b) The motor speed remains nearly constant at first – it cannot change quickly because of the inertia of the armature and the load.

(c) The suddenly-reduced counter-e.m.f. allows a greater inrush of armature current. I_a. (Armature resistance, R_a, is very low, so that the main limiting factor of armature current in a running motor is the counter-e.m.f. That is, the self-generated voltage of an operating DC machine is polarized so as to *oppose* the applied armature voltage.)

(d) As a corollary to step (c) it turns out that a *small* change in the field flux suffices to produce a *large* change in the armature current. This is a result of the arithmetic relationships – the counter-e.m.f. is normally *almost* as large as the applied armature-voltage. So, a 10% reduction in the counter-e.m.f. will cause a *much greater* percentage increase in the armature current. This important 'amplifying' mechanism will be subsequently demonstrated with numbers.

(e) The basic motor action, torque, is proportional to armature current. Thus the increase in armature current described in (d) urges the motor to speed up. Because the motor's inertia does not allow instant response, an initial runaway situation of armature current, then speed sets in. The motor *seeks* but cannot attain a condition of equilibrium. This is tantamount to describing the operating mode as one of *acceleration* in which armature current and speed continually chase one another. In plain language, we have a destructively racing motor.

(f) Summary: Opening the shunt field current of a running motor *weakens*, but does not eliminate the magnetic field from the poles. This imparts more than a mere increase in motor speed. Rather, a regenerative action sets in in which both the armature current and the speed tend to run away. Such instability can be damaging and dangerous in integral horsepower DC motors. With the small fractional horsepower motors commonly used in electronics, there may not be much hazard involved, but the malperformance can be troubling and puzzling. With the larger motors, it often happens that the resistance of the DC feeder-lines provides limitation to everything going to infinity. Similarly, fuses and circuit breakers in the line can provide protection against the more disastrous consequences of runaway operation.

Here's the alluded example of the 'arithmetic amplification' mentioned in step (d).

The equation for armature current, I_a, is $I_a = (V_a - E_c)/R_a$
where V_a is the applied armature voltage
where E_c is the counter-e.m.f.
where R_a is the armature resistance.

For our practical purposes, we can dispense with R_a inasmuch as it is both small and nearly constant; it assumes importance primarily during *starting*. Let's just assume that R_a is one ohm (Ω)

Accordingly, we will work with

$$I_a = \frac{V_a - E_c}{1}$$

For ease of mental arithmetic, let $V_a = 110$ volts (V), let $E_c = 100$ V. Then, the armature current, I_a, will be 10 amperes (A).

If, now we *reduce* the field strength, and therefore the counter-e.m.f. by just 10%, we find that the new armature current is $110 - 90$, or 20 A, an increase of 100%. Thus, a *little* change in field strength produces a similar percentage change in counter-e.m.f., but a *very much larger* change in armature current (and torque). Speed control by field current adjustment is fine, but speed instability tends to be the price of excessively weak magnetic field flux. Don't stretch the maker's specifications.

Operation of the DC servo motor

Consider now the DC servo motor. This is essentially a DC shunt motor which has been specially designed for low inertia, symmetrical rotation and smooth low-speed characteristics. These optimizations, however, do not account for the apparently strange fact that its speed versus field current behaviour is just *opposite* that of the conventional shunt motor. That is, at zero field current, the speed of the servo motor is zero. As field current is increased, the speed *increases* proportionally. Otherwise, it shares with the conventional shunt motor the ability to reverse direction of rotation with reversal of the field current. There is no speed instability tendency as in

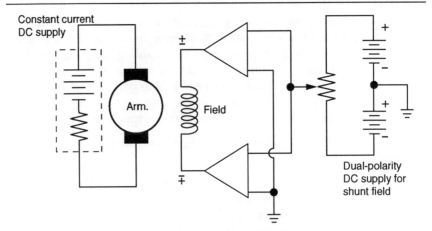

Fig. 2.7 Practical application of the DC servo motor. Although specially optimized in some of its design features, the DC servo motor is basically a DC shunt motor. Its unique response stems primarily from its method of implementation; it will be noted that the armature is fed from a constant-current DC source. This implies that V_a, the armature voltage has the freedom of variation.

conventional shunt motors. (Over-excitation of the field winding could produce out-of-rating speed, but this is not the same as the racing phenomenon of conventional shunt motors.) It is only natural to ponder the *reason* for the differences in the two shunt motors. In so doing, it will be helpful to refer to Fig. 2.7.

As stated, the radically different speed characteristic is not due to the aforementioned design differences in the servo motor. It stems from the *way* the armature is powered. Whereas conventional shunt motors derive armature current from a constant voltage (or nearly so) source, the servo motor is driven from a *constant-current* source of armature power. Although the transition in behaviour resulting from this technique may require some thought, it should be immediately clear that the armature current, now being independent of changes in field strength, is not free to engage in torque producing surges. No acceleration is developed as the field is weakened; speed is directly proportional to just the *one* quantity, field strength. The practical proof of this altered response is that a conventional DC shunt motor that is provided armature current from a constant-current source will behave essentially like the more ideal servo motor.

Having resolved this paradox, it is interesting to contemplate that the operation of the servo motor is similar to the way many people think a conventional DC shunt motor *should* behave.

We can pursue another line of thought with regard to the speed behaviour of the shunt motors already referred to provided with constant-current

Practical aspects of DC motors 49

field supplies. When the effective resistance of any kind of DC load changes, it is the *voltage* across the load that will experience much freedom to change. (Note that this is the mirror-image of the more usual situation in which load *current* changes because of a constant-voltage DC source.)

With this in mind, it should be clear that a reduction in field strength in the DC servo motor changes the armature voltage. V_a, *not* I_a the armature current. More specifically, when ϕ, the field strength is reduced, so is V_a. Although both numerator and denominator of the speed equation have been reduced, the numerator dominates because its percentage change is *greater* than that of the denominator. This follows from the arithmetical amplifying mechanism alluded to in our previous discussion. In this instance, a small decrease in ϕ brings about a much larger percentage of decrease in V_a. This being the case, the motor's speed decreases as the field is made weaker. Of *practical* significance, there are no regenerative runaway tendencies in this type of application.

In order to obtain good servo motor performance from the DC shunt motor, special considerations must be focused on certain design features. Minimization of mechanical inertia, brush friction, and windage are a must in the interest of good dynamic response. At the same time, the inductance of the field winding must be kept low so that the inductive time-constant will not be the bottle-neck that slows the response. Attention must also be directed to the torque characteristics at zero and slow speeds. Here, it is somewhat difficult to get smooth rotation. Finally it remains important to maintain good commutation over a wide range of speeds.

An alternative servo motor scheme is shown in Fig. 2.8. Direction and magnitude of motor rotation depends upon the *difference* in field strength of the two opposing windings. This greatly overcomes delay from the inductive time-constant inasmuch as the windings remain energized.

The motor characteristic of the DC watthour meter

The venerable DC watthour meter illustrated in Fig. 2.9 may be thought of as an ironless shunt motor with the shunt field current governed by the load. Interestingly, the speed response is *opposite* that of conventional DC shunt motors. Most textbooks attribute this to the absence of iron in the magnetic circuit, usually stating that this causes torque to be a direct function of the shunt field strength. This is essentially true, but it comes about in a rather indirect way.

The counter-e.m.f. generated by such an arrangement is relatively low. This is why we see the current-list resistance in the armature circuit. This resistance does more than restrict damaging current flow to the armature; it delivers constant current despite variations of counter-e.m.f. The restraint on armature current variation causes the *voltage* impressed on the armature to vary in its stead. Specifically, when there is an increase in the strength of the

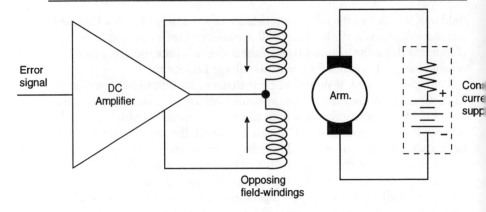

Fig. 2.8 *Split field servo motor application. This servo system makes use of a servo motor with two shunt field windings that produce opposing fields. Response is enhanced because the windings remain energized, greatly reducing the effect of their inductive time constants.*

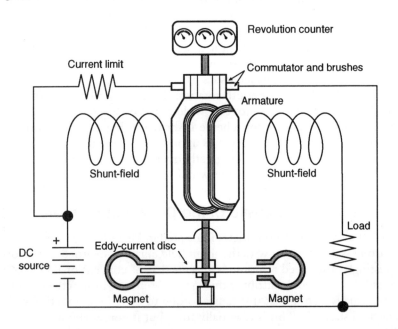

Fig. 2.9 *The DC watthourmeter – a 'contrarian' DC shunt motor. This archaic instrument is a basic shunt motor. Yet, its behaviour is opposite that of conventional DC shunt motors. Resolution of this paradox provides instructive insight into motor operation.*

shunt field, the armature voltage increases. In turn, this results in greater speed.

The eddy-current disc provides a retarding torque such that rotational speed is directly proportional to the power expended in the load. Incidentally, the armature current is *not* constant with respect to voltage variations from the DC source. If, for example, this voltage should increase, higher armature current would result. In turn, higher speed would be developed, which represents the increased power in the load. Finally, the revolution counter reads out the integrated value of this power with respect to time which is usually in kilowatt-hours of electrical energy.

The DC series motor

The DC series motor is one of the traditional 'work-horses' of electromechanical energy conversion. Its popularity with railway, automobile, and other traction applications largely stems from the very high torque developed by this motor at standstill and at low speeds. The torque is high because armature current flows through *both*, the armature and the series field giving rise to an exceptionally-strong magnetic field between the armature conductors and the pole faces. The motor obligingly supplies greater torque in response to shaft loading until a limit is imposed because of magnetic saturation. The older control method via a series-connected resistance was naturally very wasteful of energy – it sufficed for trams, but is unacceptable for electric automobiles. On the other hand, control by duty-cycle modulation of pulsed waveforms is extremely efficient and allows smooth variation of speed and torque.

One should not experiment or work with DC series motors without an awareness of a rather nasty aspect of their characteristics. It is simply that the series motor, if deprived of its mechanical load, will accelerate to very high speeds. See Fig. 2.10. Indeed, such racing can culminate in centrifugal explosion of the armature and it would be hazardous for an operator to be anywhere close to such a self-destroying machine. One might say that the unloaded motor develops too much torque for its own good due to the multiplying effect of the internal field being the product of *both* the armature and series field current. This abundant torque manifests itself as speed acceleration. At some extremely high speed, the unloaded motor would develop enough counter-e.m.f. so that armature and field current would again be relatively low; the acceleration would then come to a halt. Unfortunately, before such a new equilibrium can be attained, the motor could be expected to destroy itself.

The above-described malperformance pertains to integral-horsepower motors. Very small fractional-horsepower series motors are not likely to suffer in this fashion. This is because the bearing friction, windage, eddy currents and hysteresis all serve to limit the ultimate speed of the accelerating

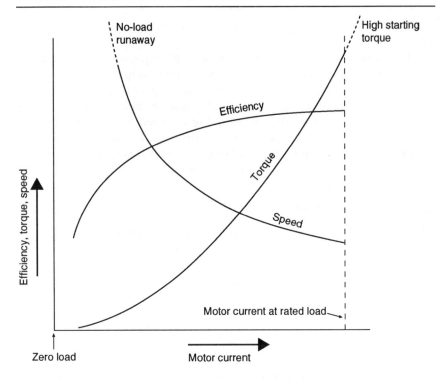

Fig. 2.10 *General characteristics of the DC series motor. In order to measure the indicated parameters, the motor current is adjusted by varying the mechanical loading applied to the motor shaft. The salient features of DC series motors are: (a) high torque at standstill and low speeds, (b) poor speed regulation, (c) runaway tendency at no-load, (d) reversal by transposing either armature or field connections, (e) basically operative from AC.*

armature. However, the general racing phenomenon will still be experienced at no load.

Reversing the rotation of a series motor cannot be brought about by simply reversing the polarity of the applied voltage. Doing so will reverse the current flow in both armature and field and the net result will be that torque will continue to be exerted in the same direction. Reversing current flow in *either* the armature or in the series field winding will do the trick. It may not always be convenient to get at the two leads of the series winding. Obligingly, motor manufacturers make two-winding series motors. Using one of the windings enables rotation in one direction. Using the other winding results in the opposite direction of rotation. This can also be accomplished with a centre-tapped field winding. A nice thing about such a provision is that one can be certain that the motor will not exhibit a favoured direction of rotation because of commutation problems or because of the physical orientation of the brushes.

An interesting consequence of the refusal of the series motor to change its direction of rotation in response to reversal of the applied voltage polarity is that such a motor will operate on alternating current as well as direct current. This does not imply that any DC series motor will perform as satisfactorily on AC as on DC. Usually, some design modifications are required if the motor is intended for AC operation. The motor designer must give special attention to eddy-currents hysteresis, inductances, etc. When thus optimized for both AC and DC operation, the series motor is known as a universal motor. In actual practice, it will likely be found that the universal motor is not as effective on DC as a series motor designed for DC operation.

The starter motor of automobiles is a noteworthy application of the series motor. Here, tremendous torque for a relatively brief period is needed to turn over a cold engine. On the other hand, the AC motors in vacuum cleaners are usually universal types. High starting torque is not required, but high speed is needed to make these appliances effective. At the same time, the fan imposes sufficient loading to prevent destructive racing.

The DC compound motor

The DC compound motor has both series-connected and shunt-connected field windings. By varying the polarization and the current in these field windings, the motor can be made to behave as a series motor, a shunt motor, or a motor exhibiting various blends of series and shunt motor characteristics. Significantly, it can be made to eliminate the no-field runaway characteristic of the shunt motor and the no-load runaway characteristic of the series motor. The compound motor can be an ideal machine for the experimenter and for the practical engineer seeking to optimize certain behaviour. The current in the series-field winding can be diverted by means of a resistance connected across its terminals. Control of the current in the shunt-field winding is achieved by a series-connected rheostat, just as in a 'plain vanilla' shunt-motor. The speed–load behaviour of compound motors relative to the other DC motors is shown in Fig. 2.11.

In the cumulatively-compounded motor, the series field *aids* the shunt field. In the differentially-compounded motor, the series field *opposes* the shunt field. In both situations, the flux relationship between the magnetic field of the armature and the net magnetic-field from the windings on the poles determines the direction of rotation. Additionally, there are two connection schemes which can 'fine-tune' the motor characteristics, but do not in themselves determine whether one has a cumulatively or differentially-compounded machine. These are the so-called long-shunt and the short-shunt connections shown in Fig. 2.12. Ordinarily, the choice between the two is dictated by wiring convenience. As mentioned elsewhere, it is very easy to get into trouble if a rotation-reversing switch is needed.

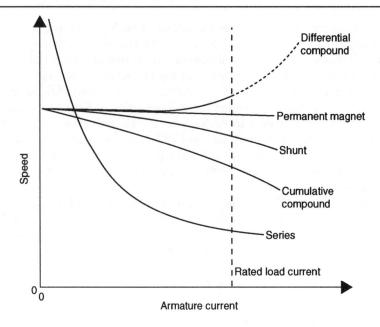

Fig. 2.11 *The effect of load on speed in DC motors. Armature current is a good representative of mechanical loading, which is difficult to measure and record. Runaway is possible not only in the* unloaded *series motor, but in the* fully-loaded *differentially compounded motor. The other motors are generally free of such load-sensitive instabilities. The permanent magnet type is best choice for good speed regulation.*

Care should be exercised to connect this switch as depicted in the two circuits of Fig. 2.12.

Do not attempt to convert a permanent magnet motor to a compound motor by placing a few turns of heavy wire over the permanent magnet structure. Initial success with this technique is likely to be short-lived because of demagnetization of the permanent magnet.

Motor and generator performance in the same machine

Hobbyists are often confronted with a situation in which an available machine needed to function as a generator is actually a DC motor. Conversely a DC generator may serve a purpose if it can be used as a motor. It is well known that motor and generator action always occur simultaneously, but the practical question is whether a given machine can be freely used one way or the other. Some engineering texts lead one to believe that, in some cases at least, there is perfect interchangeability between the functions – it is as if the manufacturer labels the same machine either as a motor or a generator depending on market demand.

Practical aspects of DC motors 55

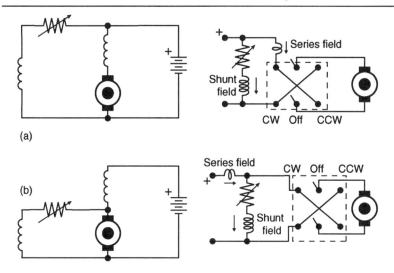

Fig. 2.12 *Either of two connection schemes can be used in compound DC motors. The two arrangements yield approximately the same motor performance. However, care must be exercised in providing for the reversal of rotation.*
(a) The long-shunt *connection and its method for reversal.*
(b) The short-shunt *connection and its method for reversal.*

Other texts emphasize the difficulties and malperformance one can expect from forcing a machine to develop the function not intended by its designers. In actual practice, individual applications need to be evaluated to determine feasibility. Quite often a less than optimum performance can nevertheless prove satisfactory. Nor is it uncommon to find that certain machines prove equally happy performing as motors or as generators. Often too, the experimenter can impart some minor surgery to tame an otherwise-misbehaving machine.

If, indeed, trouble is encountered from an attempt to interchange a machine's intended function, it will likely stem from the effects of *armature reaction*. This is a form of air-gap flux distortion which twists the magnetic neutral axis so that the physical location of the brushes is no longer where it should be. This adversely affects development of both torque and counter-e.m.f., but the most objectional malperformance in both motors and generators is sparking and arcing at the commutator, that is, bad commutation.

The knotty feature of armature reaction is that its effect is *opposite* in the motor and the generator. In the motor, shift of the neutral axis is *backward* to the direction of rotation. In the generator, this shift is *forward* of the direction of rotation. Interchange of function is, therefore, *doubly* plagued by the deleterious effects of armature reaction if the brushes had initially been positioned to favour one function. Fortunately, this worst-case situation is not often encountered.

The preceding paragraph was premised on machines constructed so that the brush assembly can be manually rocked throughout an angle in order to find the neutral magnetic axis displaced by armature reaction. Large machines were at one time made this way; some fractional horsepower motors and generators incorporated this technique. Modern machines tend to use more sophisticated methods for dealing with the effects of armature reaction. Attention given to the shape and fabrication of the pole-pieces, choice of magnetic materials, a small air-gap, and a strong main field can go a long way in making a machine less susceptible to armature reaction. With a little compromise added to the mix, DC motors and generators can be designed that can readily be used for either function.

Small salient interpoles installed between the main poles can be employed to more or less effectively neutralize the tendency of armature reaction to shift the neutral axis of the main field. These are actually called *interpoles* and comprise small windings all connected in series and connected in series with the armature. Schematically, such an arrangement is suggestive of a compound motor, but the positioning and operation of these interpoles produces negligible compounding effect. In practice, it is necessary to keep in mind that the connection of the interpoles to the armature must be *reversed* when changing machine operation from say, a motor to a generator. Otherwise commutation difficulties will be agitated, not alleviated.

Large machines can also make use of so-called *compensating windings*. These are embedded in the pole-faces and, at least superficially, resemble the damper windings used in AC synchronous motors. These compensating windings also counteract the shift of the neutral axis that the armature reaction tends to cause. The compensating windings are also connected in series with the armature and require reversal when changing the function of the machine, say from motor to generator.

It is to be noted that changing the polarity of a generator or the direction of a motor's rotation doesn't require the minor surgeries alluded to. In any event, whether a machine's function can be interchanged often depends on the hobbyist's or the engineer's viewpoint.

Reversing the rotation of DC motors

It was pointed out that reversal of rotation could be achieved in both series and shunt motors by reversing either the field or the armature connections, but *not* both. Reversing both would leave the present direction of rotation unchanged. In the series motor, it is often convenient to reverse the field winding connections. In the shunt motor, it is often felt that reversing the current flow in the shunt field is the wise thing to do because one only has to switch relatively-low currents in the shunt field winding. This, however, may not be the best logic.

The shunt field winding is highly inductive compared to the armature. This can result in destructive *arcing* during rotation switching. Unless action is taken to suppress or absorb the arcing energy, switch replacement is likely to become a regular feature on the maintenance schedule. The arcing can also be a source of EMI and RFI. It is probably better to do the polarity switching in the *armature* circuit as, despite the relatively heavy current, arcing is likely to be minimal because of the low inductance of the armature. One must also be mindful that when a switch in the shunt field circuit becomes badly damaged from the repeated arcing, the field current path may be *opened*, causing the motor to go into its destructive mode of racing. As described elsewhere, this involves acceleration of both speed and armature current. The runaway operation can damage the motor and endanger people. In contrast, nothing drastic is likely to happen because of faulty armature switching. (Short-circuiting is far less likely than failure of continuity.)

Another practical fact of operation involves *compound* motors. Here the clear and clean way to reverse rotation is to switch the polarity of armature current. An attempt, unless carefully implemented, to switch one of the field currents can easily result in differentially-compounded operation in one direction and cumulatively-compounded operation in the other direction. Such malperformance could obviously be confusing. It is usually safe to reverse rotation during operation at rated speed and load. This is known as *plugging*. It quickly brings the motor to standstill just prior to reversal.

Practical use of counter-e.m.f.

A practical example of the use of a motor's counter-e.m.f. is evidenced in the speed-control circuit of Fig. 2.13. Not only does this simple circuit enable the speed of a universal or a series motor to be varied over a wide range, but the inherent speed-regulation causes the otherwise-wild speed–load relationship of these motors to be stabilized. In other words, at any selected speed setting, the motor will operate at near-constant speed despite variation of the load. The mere mention of the term 'regulation' suggests the involvement of a negative feedback-loop in the corrective action. Yet, we see neither a feedback path nor the reference voltage it would be expected to be associated with. How, indeed, is the speed of the motor restored to a set value?

Assume that the motor is operating at some set speed and is subjected to an increased mechanical load. The natural impulse of the motor is to slow down. However, the result of such reduced speed would be a reduction in its counter-e.m.f., recalling that motor and generator action occur simultaneously. Between the current pulses delivered by the SCR, the coasting

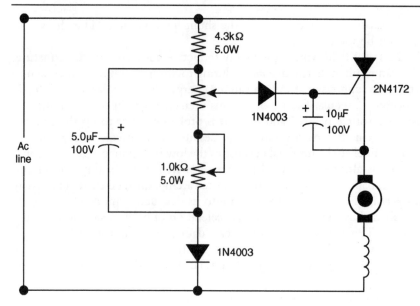

Fig. 2.13 *A practical use of the counter-e.m.f. of a motor. The* generator *action of the motor is sensed by the gate of the SCR. Speed-up of the motor causes increased delay in the triggering of the SCR. In turn, this allows less motor current, less torque and therefore reduced speed. The converse sequence occurs if the motor attempts to slow down. Thus, speed regulation ensues despite changes in the mechanical loading of the motor.*

motor is actually operating as a *generator*. Only inertia keeps the load rotating. Let's see now what happens.

The SCR is enabled to fire *earlier* in the positive half-cycle of the power line voltage wave. This is because the counter-e.m.f. is so polarized as to oppose the SCR gate-trigger voltage – the less counter-e.m.f. the *easier* it is to trigger the SCR. It is the nature of the SCR that the less delay involved in its triggering, the greater is the rms value of current delivered to the electrical load. Such an increase in motor current causes it to develop more torque, which is manifested as higher speed. Inasmuch as the converse sequence of events occur if relaxation of the mechanical load impels it to speed up, it can be appreciated that the motor is under the influence of speed-regulation. Although 'invisible', feedback occurs via the counter-e.m.f.

Although this particular implementation is not very efficient, it has been used for sewing machines, mixers and similar small-appliances. The experimenter can readily adapt the basic principle for use with Triacs and with other types of motors.

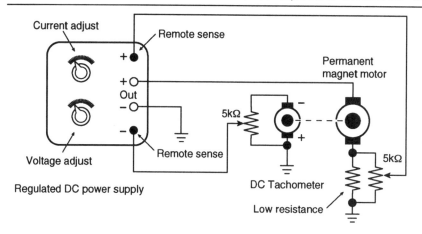

Fig. 2.14 *Speed and/or torque control of a DC permanent magnet motor. This simple arrangement demonstrates independent control of speed and torque of a permanent magnet DC motor. (A shunt motor with a separate field current supply would also be applicable.) In the interest of efficiency, the sampling resistance in the armature circuit of the PM motor should be as low as possible.*

Flexible control of permanent magnet motors

The experimental set-up shown in Fig. 2.14 is instructional and potentially-useful for a variety of electric motor applications; moreover, it can be quickly arranged at minimal cost. Briefly, it enables *independent* control of speed and torque of a permanent magnet DC motor. By recording data obtained with such a system, one is then in a position to scale up or down the important parameters that might be involved in a practical application. What makes this experimental set-up unique is that most control circuits simultaneously change *both*, the speed and the torque, disallowing the 'luxury' of controlling just one or the other.

The heart of this control technique is a source of DC power which, in itself, enables operation as a constant-voltage supply, or a constant-current supply, or as both in selectable combinations. As luck would have it, such regulated DC supplies have become commonplace not only in the engineering laboratory, but among hobbyists as well. It should be fairly easy to find one with a several to several-tens of amperes capability, and the acquisition of a small permanent magnet motor of almost any voltage rating should involve little strain of effort or pocketbook.

The guiding principle here is that the torque developed by a simple PM motor is governed by armature *current*. On the other hand, speed is a function of the *voltage* impressed on the armature. Because of these relation-

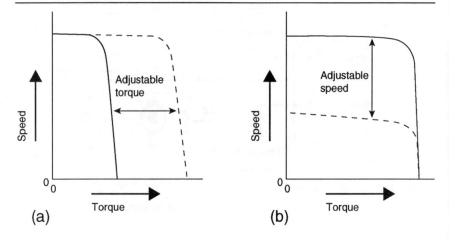

Fig. 2.15 *Independent speed and torque adjustments of permanent magnet motor.*
(a) Torque control by variation of constant-current level in the armature. Note that speed remains nearly constant.
(b) Speed control by variation of constant-voltage level applied to the armature. Note that torque remains nearly constant.

ships, it is easy to see why most control circuits change *both* speed and torque together. With the independent adjustment of these performance characteristics provided with the right kind of regulated DC supply, we have an excellent example of electronic 'tailoring' of motor behaviour. (Interestingly, the DC tachometer is a small version of the PM motor. The tachometer functions as a DC generator, delivering a voltage proportional to speed.) The speed–torque behaviour of the PM motor controlled by the regulated DC power supply is shown in Fig. 2.15.

The grey area of DC and AC motors

The separate study of DC and AC motors has long proved a practical expedient. It should be appreciated, however, that such a classification is an arbitrary one. It has already been mentioned that the armature conductors of the classic DC motors actually carry alternating current. Indeed, there have long been specialized dynamos with both commutators *and* slip-rings associated with the armature. In such machines, simultaneous operation as a DC motor and an alternator is possible. Conversely, such a dual-function machine can operate as an AC motor and as a DC generator. The practical existence of such converters is evidence that the transition from DC to AC motors is a very natural one; a conductor experiencing electromagnetic torque might well 'see' similar phenomena in both types of motors.

It is also true that some motors operate in a 'grey' area, partaking of both

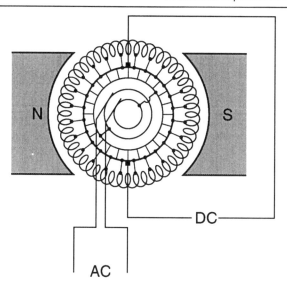

Fig. 2.16 *The synchronous converter – an AC/DC machine. It will be noted that both, a brush-commutator system and slip-rings are associated with the armature. This enables the one machine to operate either as an AC motor and DC generator, or as a DC motor and alternator. Advantage is taken of the phenomenon existing in DC motors, that there is alternating current in the armature conductors. Thus, a single machine can displace a motor–generator set.*

DC and AC formats. For example, reference to the three-phase DC brushless motor is suggestive of an oxymoron – how can one have a polyphase DC circuit? In actuality, the three main groups of stator windings in such motors are symmetrically-separated by 120°, just as in a three-phase induction motor stator. At the same time, the current pulses supplied to these windings are unidirectional and are timed to energize the windings sequentially. In other words, we have commutation and the production of a rotating magnetic field, phenomena suggestive of *both* DC and AC motors.

The above discussion can also apply to stepping motors. One manufacturer markets a stepping-type motor which can either be DC-pulsed to predictable positions, or can be continuously rotated in the manner of an AC synchronous motor. Finally, electronic control of motors as manifested by chopping techniques, pulse-width modulation, and phase-control tend to obscure the sharp demarcation between DC and AC systems. With these matters in mind, let's now look at the machines known as AC motors.

3 Practical aspects of AC motors

Around the time that alternating-current generation and transmission began to show clear evidence of ultimately replacing most direct-current systems, the alternating-current motor was fortuitously ready for widespread industrial use. This is not altogether surprising, for most alternating-current motors are basically simpler dynamos than their direct-current counterparts. This stems largely from the elimination of the brush–commutator mechanism. The slip-rings used on certain AC motors are of relatively simple nature and involve much less cost or maintenance than do brush–commutator systems. Moreover, the duty-cycle and magnitude of the slip-ring current is usually low compared to the line current.

The invention and development of the polyphase induction motor comprised another quantum leap in motor technology, for here was a self-starting motor that required no centrifugal switches and, in its most popular format, no slip-rings either. And pound for pound of iron and copper, it was a more efficient converter of electrical to mechanical energy than single-phase versions.

Although constructionally simpler than DC motors the design and sometimes the operation of AC motors can be a bit more involved than that pertaining to DC motors. We must now pay attention to inductance, skin-effect, power-factor hysteresis and eddy-currents in ways that require a higher priority than with DC motors. Our technical vocabularies must be expanded to embrace the concepts implied by slip, synchronism, rotating field, phase sequence and a few others which pave the way to the understanding of polyphase circuit behaviour. However, we shall find that electromagnetic torque derives from interplay of the same forces that work in the DC motor.

The great induction motor dilemma

Inspection of the rotor of a squirrel-cage induction motor leads to the impression that every effort has been made to endow the conducting bars with exceedingly-low resistance. They are of very heavy gauge compared to the other windings that are generally seen in motor elements. It could be quite natural to extrapolate the notion that zero-resistance conductors would be used for the squirrel-cage bars if this could be practically realized. Such a thought might be reinforced by the knowledge that low resistance squirrel-cages allow the induction motor to more closely approach synchronous operating speed. As a pay-off also, the speed regulation improves with lower resistance cage bars.

The practical fact, however, is that such an induction motor would not run at all. This conclusion derives from basic principles. For one thing, the slip of the rotor is proportional to its resistance. Accordingly, a motor with zero-resistance bars would 'want' to run at the same speed as the rotating magnetic field – that is, at synchronous speed. If, however, it did this *no* torque would be developed. Torque is the consequence of Lenz's law, which tells us that the magnetic field of the rotor eddy-currents oppose the magnetic fields inducing these currents. But, no eddy-current induction takes place unless there is a speed *difference* (slip) between the rotation of the rotor and the rotation of the magnetic field from the stator.

This can be confirmed from another viewpoint: zero-resistance conductors do not allow penetration of time-varying magnetic fields. Here again no eddy-currents would be induced and the accompanying magnetic field needed to oppose the stator field would be absent.

The resolution of our dilemma is simply that lower squirrel-cage resistance improves speed regulation and operating efficiency (at the expense of torque) but only to a point. If we had access to conductive material, say ten-times better than copper, we would find ourselves in the realm of diminishing returns. Indeed, we would begin to see unacceptable loss of torque, that is, basic motor action. Theory and fact nicely converge here.

Practical aspects of the single-phase induction motor

A 'pure' single-phase induction motor would qualify as the simplest machine for conversion of electrical to mechanical energy. Such a motor would comprise a single-winding stator and a number of short-circuited bars for the squirrel-cage rotor. Such a design has little likelihood of being encountered for the very practical reason that it is not self-starting. Note that what we would have would amount to a 'rotary transformer' with the secondary free to turn. And, indeed, it will turn providing an auxiliary starting technique is used. Moreover, such a motor will accelerate up to maximum

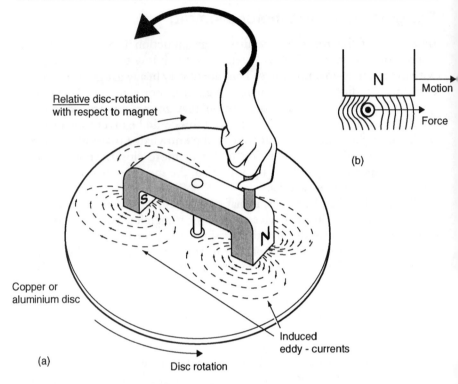

Fig. 3.1 *The principle of the induction motor. (a) The disc follows, but never quite attains the rotational speed of the rotating magnetic field. This results from the interaction of the magnetic field from the induced eddy-currents and the rotating magnetic field. This may not be intuitively obvious, but is more easily grasped by dealing with relative motions; imagine the field from the magnet to be stationary and the conductor to be moving in the opposite direction. Then, by applying Lenz's law which states that the field from the induced eddy-currents must oppose the motion producing them, cause and effect can be more easily perceived. (b) The situation beneath the poles emulates the 'motor-action' of a current-carrying conductor in a magnetic field.*

speed if the shaft is given an initial spin in either direction. The principle of the induction motor is depicted in Fig. 3.1.

This *basic* nature of the single-phase induction motor merits an explanation owing to the widespread use of this AC motor wherein it is modified by the addition of some type of starting provision. If the stator of the basic induction motor is supplied with an AC voltage, the short-circuited secondary will be subjected to a pulsating magnetic field comprised of the interaction between the stator field and the field from the induced current in the squirrel-cage shorted-secondary. Nothing will happen because torque will alternately be developed in one direction and then the other. The net torque

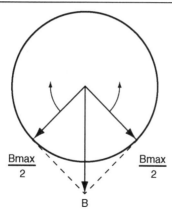

Fig. 3.2 *Standstill torque in the single-phase induction motor. Apparently the esoteric approach of the mathematician, the vector diagram of the torque situation has practical significance. The single-phase pulsating torque B can be equivalently represented by two counter-rotating torques of half-amplitude. This tells us that the rotor remains stationary not because it feels no torque, but rather because it is alternately imparted torque in both directions. Once the rotor is rotated in either direction the balance between the two competing torques is upset and acceleration occurs.*

over any number of alternating-current cycles will average to zero.

Interestingly, the pulsating field can be mathematically analysed to consist of two oppositely-rotating fields. This turns out to be practically true, for if the motor is somehow given an 'artificial' start, the balance between the oppositely-rotating fields will be upset and the rotor will begin to follow one or the other. At first, the slip speed will be just under 100%, but very quickly this initial rotation will be accelerated to full speed, perhaps just a few per cent below the synchronous speed. The standstill torque situation is shown in Fig. 3.2.

The salient operating feature is that rotational torque sets in as soon as slip is less than 100%. We should also note that torque would again become zero *if* the motor reached synchronous speed and therefore, induction motors run slightly *below* the speed of the rotating field from the stator.

Split-phase starting techniques for induction motors

Early motor experimenters observed that a 'textbook-pure' rotating electromagnetic field was not necessary to cause a single-phase induction motor to be self-starting. in other words, a phase-splitting technique that produced two AC voltages electrically displaced somewhat *less* than the ideal 90° would do the trick. This is not surprising inasmuch as capacitors, especially electrolytic types, may depart considerably from ideal characteristics and fail to provide a full 90° of phase shift.

The practical outcome was a so-called 'split-phase' induction motor with two 90° space-displaced windings as with the capacitor motor. However, in this case, the windings are *not* identical; one of them – the 'running' winding – has many turns of heavy wire whereas the 'starting' winding has fewer turns of smaller-gauge wire. Therefore, the running winding has relatively high reactance and low resistance compared to the values of these parameters in the starting winding.

It turns out that if these windings are connected in parallel, there is considerable phase shift between their currents, and therefore their electromagnetic fields. Thus, the squirrel-cage rotor finds itself in a reasonable facsimile of a two-phase rotating field and starts accelerating from a standstill.

Actually, practical motors of this kind are also equipped with centrifugal switches which disconnect the starting winding when the speed is fairly close to its rated value at the rated load. Otherwise, the high I^2R dissipation in this winding would lower the operating efficiency and would create sufficient heat to damage motor insulation. This works out well for, once the starting winding has performed its function, its contribution to motor torque is negligible.

Reversing either winding will reverse the rotation. However, the motor must be brought to standstill first. It will not respond to 'plugging', i.e., to reversal while still rotating in one direction. This remains true even if the centrifugal switch is bypassed.

Types of single-phase AC motors

The basic DC motors are the shunt, permanent magnet, series several compound types, which blend the characteristics of shunt and series motors. Selection among these enables various speed–torque relationships to be realized. At first consideration, it would appear that life would be simpler with AC motors where one only needed to choose between the induction and the synchronous motors.

It happens, however, that neither of these AC machines are self-starting on a single-phase AC line. Hence, many techniques have been incorporated to cause these motors to accelerate from a standstill to their normal operating speeds. The practical ramification of this is that such AC motors are not quite as simple as they might be were they inherently self-starting. Starting schemes are shown in Fig. 3.3.

Inasmuch as a polyphase induction motor *is* self-starting, it was only natural to make the single-phase motors 'think' they were polyphase types operating from a polyphase line. This, alone, gave birth to a number of different induction motors. The easiest and most direct way of accomplishing this is to provide two windings on the stator. These are displaced in *space* 90° with respect to each other. It is also necessary to 'split' the single-phase AC line so that the motor 'sees' *two* voltage sources, with one displaced from

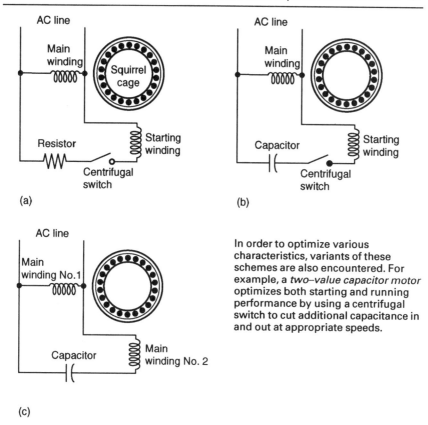

Fig. 3.3 *Single-phase induction motors and their starting arrangements. Ideally, the fields from the two windings should be displaced from one another by 90° in both space and time. Practically, this is not attained. The centrifugal switch opens the starting windings at about 80% of synchronous speed.*
(a) Resistance-start, split-phase motor. The physical resistance need not be present and instead the resistance can be incorporated in the starting winding.
(b) Capacitor-start motor. Here capacitance is used to produce greater phase displacement between the windings.
(c) Permanent split-capacitor motor. The two windings are identical and since both remain in the circuit, no centrifugal switch is used. This is essentially a two-phase motor which operates from the single-phase line.

the other by 90 *electrical* degrees. In other words, we wind up with a two-phase motor powered from a two-phase AC source. Such a single-phase motor operates as though it were a true two-phase motor energized from a true two-phase source. Not only does it develop starting torque but its running characteristics can be exceptionally smooth and efficient.

The natural phase-splitter is a *capacitor*. Because of cost, size, and reliability

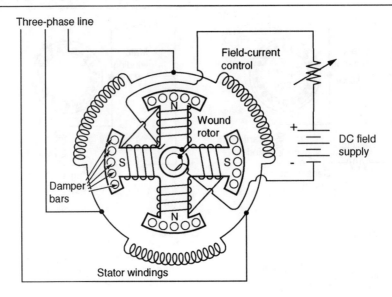

Fig. 3.4 *Basic essentials of the three-phase wound-rotor synchronous motor. The wound-rotor is connected to the external DC source through slip-rings. The damper windings are actually short-circuited bars which simulate the squirrel-cage structure of induction motors. They enable self-starting as an induction motor.*

factors, there are a variety of capacitor-type split-phase induction motors. Some open the capacitor circuit with a centrifugal switch once the motor has started. In contrast to such a 'capacitor-start' motor, some types known as 'capacitor-run' motors retain the phase-splitting capacitor. With this type, the stator may be nearly identical to an actual two-phase motor — one designed to operate from a *true* two-phase source. In small sizes, this type is very useful as an AC servo motor and is readily controlled in electronic systems via the drive to one of its two phase-windings.

The synchronous motor

The classic synchronous motor depicted in Fig. 3.4 is an extraordinarily interesting dynamo. It long pre-dates the electronics era, but has nonetheless had unique applications aside from its torque-providing ability. Although its history has long been associated with multi-integral-horsepower ratings comparable with the large induction motors used in industry, sub-fractional horsepower versions now offer many potential uses in conjunction with solid-state IC controllers. A representative rotor is shown in Fig. 3.5.

The unique aspect of the wound-rotor synchronous motor is that it can behave as a *variable reactance*. As such, it can simulate a usefully-wide range of

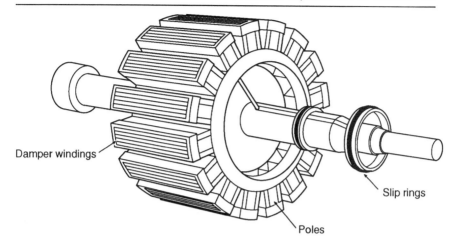

Fig. 3.5 *The rotor of a three-phase synchronous motor. Without the damper windings in the pole-faces, the synchronous motor would not be self-starting. They also discourage hunting behaviour which might be provoked by sudden changes in the load or the applied voltage. Both the rotor and the stator are essentially identical to those of alternators. Note the general similarity to the automobile alternator which, however, has no need for the damper windings.*

both inductive reactance and capacitive reactance. Historically, it has been much used to remedy the poor power factor presented to the utility power lines by the popular induction motors. Lightly or partially-loaded induction motors appear inductive and cause the line current to lag behind the line voltage. This means that the AC transmission lines have to carry more current than would be necessary at unity power factor. In a large system the alternators at the power station must then provide this extra load. There are associated side-effects too; for instance, line voltage regulation is severely degraded and circuit-breakers and fuses no longer behave dependably. It makes sense that there are often penalties to factories operating with low power factors.

One might suggest a physical capacitor connected across the line to neutralize the inductive reactance and thereby improve the power factor. Indeed, this technique has been used. However, the size and cost of such capacitors quickly becomes prohibitive in large installations and tapped switches for selecting optimum capacitance for the prevailing load involves a lot of 'monkey business', and again *cost*.

A more elegant solution has been provided by simply installing a synchronous motor of appropriate kVA rating. The 'elegance' stems from the fact that such a motor can *also* provide valuable mechanical horsepower and, at the same time, the manual or automatic control of relatively low DC current to its rotor can improve the overall power factor by making the

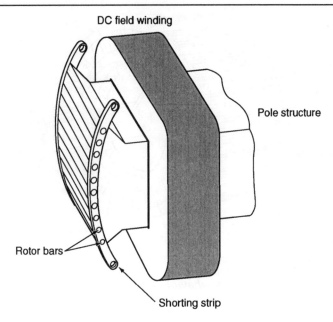

Fig. 3.6 *Damper windings enable the polyphase synchronous motor to self-start. These pole-face 'windings', also known as* amortisseur *windings, emulate the function of the squirrel-cage structure in the induction motor. Thus, a three-phase synchronous motor self-starts as an* induction *motor and it then accelerates until its speed is within a few per cent of the speed of the rotating electro-magnetic field of the stator. Finally, by way of an incremental jump, actual synchronous-speed is attained. Damper windings also discourage the 'hunting' tendency of the rotor. Damper windings cannot develop starting torque for single-phase synchronous motors.*

power line 'think' it is delivering current to a load more resistive in nature than the load imposed by inductive motors alone.

It seems a paradox that the polyphase synchronous motor cannot, of its own accord, develop a starting torque. The two- and three-phase *induction* motors are self-starters. Three-phase induction motors and three-phase synchronous motors use essentially the same stators. In both instances, these stators develop rotating electromagnetic-fields; yet the unencumbered rotor of a synchronous motor feels *no* rotational force when at a standstill. In order to self-start, the squirrel-cage structures already alluded to as damper windings have to be recessed in the pole faces. A detailed view of these structures is shown in Fig. 3.6.

Because of the damper windings, the polyphase synchronous motor self-starts as an *induction* motor. It accelerates to a speed just a few per cent under the speed of the stator-supplied rotating field and then the rotor jumps slightly ahead and locks into synchronism. Thereafter it continues to rotate at synchronous speed. If more load is imposed on the motor the rotor slips

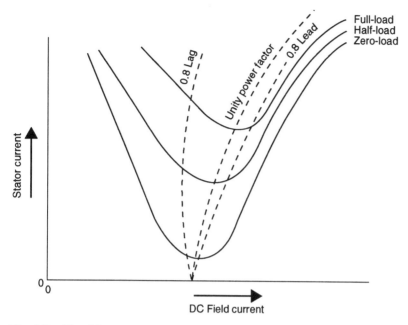

Fig. 3.7 *The 'V' curves of the wound-rotor synchronous motor. Leading or lagging power factor can be presented to the power line by merely adjusting the field current. Not only does the motor remain in synchronism, but as shown, it can simultaneously drive a load.*

back a bit but its *average* speed remains synchronous with the rotating magnetic field from the stator. If too much load is imposed, breakdown torque will be exceeded and the motor will come to a halt. The motor runs *at* synchronism or *not at all*.

A natural question that arises is why the overloaded synchronous motor doesn't just drop back in speed and continue to run as an induction motor. There are two reasons; first, the damper-winding torque is not sufficient for the load and second, a synchronous motor with an energized field winding develops no torque – that is why dependency for starting became the burden of the damper-windings originally. Torque is not otherwise developed in the rotor unless it is locked into synchronism.

The allusion above to the slipping back of the rotor in response to loading is not to be confused with the concept of *slip* in a motor operating on the induction principle. In the synchronous motor, the backward slip is *momentary* and the rotor continues to operate at synchronous speed. In the induction motor, there is an actual *reduction* in rotational speed.

The unique 'resonance' curves of the synchronous motor are illustrated in Fig. 3.7. Note the resemblance to *LC* 'tank' behaviour. Here, the tuning is done by adjusting the DC field current. It can be seen that unity power factor operation can be had for any load situation. This contrasts with induction motors where the lagging power factor becomes progressively worse as the load is lightened. A nice feature of this wound-rotor synchronous motor is that it maintains its synchronous speed over wide ranges of field current and load. Indeed, it will *only* operate at synchronous speed; if ultimately forced to deliver more than its maximum, or pull-out, torque, it immediately comes to a standstill.

From a practical standpoint, it is easy to achieve unity-power factor operation or very close to it, by simply 'tuning' for minimum stator current. Quite often, however, one would be more likely to over-excite the field in order to cause the line current to *lead* the line voltage. As already pointed out, this enables the motor to act as a capacitor and provide compensation for the lagging power factor usually imposed by other electrical loads such as the alluded induction motors. In this way, the net power factor of the whole group of loads at a given locality can be greatly improved. In practice, it would be very desirable to improve a 0.65 power factor, but there would be a diminishing return in attempting further improvement of a 0.9 power factor. In any event, the wound-rotor synchronous motor obligingly provides its mechanical horsepower while simultaneously upgrading the power factor of the line.

Two important parameters of operation are the pull-in torque and the pull-out torque. As its name implies, pull-in torque just enables the motor to lock into synchronism as if more torque were demanded, the motor would not attain synchronous speed. One can also deal with a minimum speed needed to make the final jump into synchronous operation. On the other hand if progressively more torque is demanded from the already-operating motor, the pull-out torque will be reached whereupon the motor will abruptly come to a standstill. Pull-in and pull-out torque are depicted in the performance curves of Fig. 3.8. In practice, interaction between these torque levels and the field current tends to be small enough so as not to be bothersome.

Shaded-pole motors

The shaded-pole induction and synchronous motors represent a neat technique for imparting starting-torque to such single-phase AC motors. The starting torque developed by shading-pole design is quite low compared to that available from universal motors, repulsion motors, and the various split-phase types; this tends to limit the shading-pole motors to about 1/20 horsepower. On the other hand, when one gets away from heavy industry it is observable that the sub-fractional electric motor is widely applied for diverse purposes. For this reason, shaded-pole motors, which dispense with brushes, commutators, slip-rings and centrifugal switches, have proven to be exceptionally reliable and cost-effective.

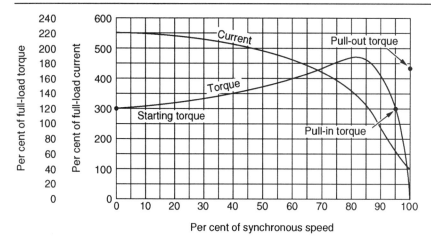

Fig. 3.8 *Important operating features of the wound-rotor synchronous motor. Until 100% of synchronous speed is attained, the motor runs as an induction motor. Note the high starting-torque.*

At high speed, providing pull-in *torque demand is not exceeded, the motor locks into synchronous speed. If a higher torque level known as the* pull-out *torque is exceeded, the motor will abruptly fall out of synchronism and come to a standstill.*

These motors make use of high-conductance copper rings fitted snugly on the edgewise portions of the pole structure, as can be seen in Figs. 3.9 and 3.10. These rings, or shading-coils, act as short-circuited secondaries of the stator winding. The magnetic fields produced by the currents induced in the shading-coils incur a delay with respect to the magnetic field developed in the major face of the pole structure. This delay, or out-of-phase condition gives rise to a sweeping action of magnetic flux in the air gap. Although this is not a pure rotating field, it does serve to impart uni-directional torque to the rotor.

In order to reverse shaded-pole motors, the 'shoe needs to be on the other foot' i.e., the shading-coil would have to be associated with the opposite edge of the pole-pieces. This is indicated as positions X and Y in Fig. 3.9. Obviously, most of these motors are not readily reversed. As might be anticipated, special reversible models have been made with switchable shading-coils on the poles. These, however, are rarely encountered in practice.

The hysteresis motor

The shaded-pole synchronous motor depicted in Fig. 3.10 is known as a *hysteresis* motor. The dissipation phenomenon of hysteresis, generally minimised in electrical design, is *beneficially* used in this motor. This concerns the

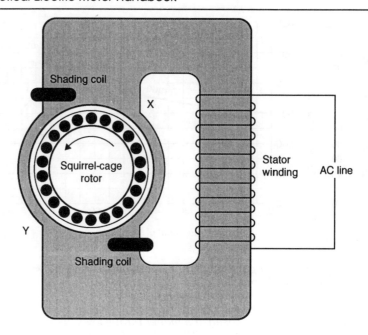

Fig. 3.9 *The basic single-phase shaded-pole motor. The shading coils are heavy-gauge copper rings and act as shorted transformer-secondaries. Their induced currents produce portions of magnetic flux out of phase with the flux produced by the main faces of the field structure. The resultant distortion of the air-gap magnetic field includes a weak rotating component. This is sufficient to develop enough starting torque to impart rotation to the squirrel-cage rotor. Acceleration then sets in until the motor runs at slightly sub-synchronous speed. Reversal would require placement of the shading coils at corners 'X' and 'Y' of the pole structure.*

rotor, for otherwise both structure and operation identify the motor as a single-phase, shaded-pole induction motor. Indeed, it has much in common with the motor of Fig. 3.9.

The rotor of the hysteresis motor is comprised of laminated cobalt-steel stampings. Because this magnetic material displays high retentivity, it also develops high hysteresis loss when subjected to alternating currents. Also relevant to the performance of this motor, this material has high electrical resistivity. Let's see how these attributes combine to endow this motor with unique performance.

To begin with, we obviously are dealing with an easily-produced and cost-effective product with predictably low maintenance. In starting from standstill, this motor behaves similarly to the squirrel-cage motor of Fig. 3.9. Starting torque tends to be high because of the resistivity of the rotor which improves its power factor. (This is essentially similar to the torque enhancement obtained in wound-rotor motors by *purposely* inserting rotor-resist-

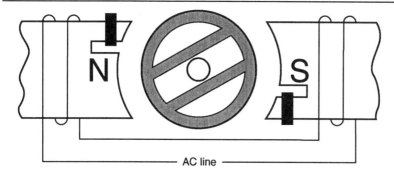

Fig. 3.10 *The hysteresis motor. These are, for the greater part, single-phase, shaded-pole, induction motors. Hysteresis loss is high in the high-retentivity steel rotor. However, it serves a good purpose in similar fashion to the resistive loss in a wound-rotor machine. The hysteresis loss increases the torque of the motor at an acceptable impairment of efficiency. Additionally, the crossbars become magnetized and cause the rotor to attain synchronous speed by locking in step with the rotating magnetic field.*

ance.) Then, as rotation commences, the hysteresis loss in the rotor also represents appreciable power dissipation. Although this adversely affects the efficiency of the motor, it continues to improve low-speed torque. All in all, this motor's ability to start and accelerate to its operating speed is fortuitously enhanced by its electrical and magnetic loss mechanisms.

Next, as synchronous speed is approached, the magnetic retentivity of the crossbars allow the development of semi-permanent poles in these elements of the rotor. For all practical purposes, the crossbars now behave as a bar magnet. This causes the rotor to make the final incremental increase in speed and lock in step with the rotating magnetic field of the shaded-pole structure. Thus, *aided* by electrical and magnetic losses, the hysteresis motor starts as an induction motor and attains operational status as a *synchronous motor*.

The reluctance motor

The reluctance motor is an interesting and useful hybrid of the squirrel-cage induction motor and the synchronous motor. Unlike ordinary synchronous motors which come to a standstill if too greatly loaded, the reluctance motor tends to merely fall back from synchronism, but continues to exert strong torque as a squirrel-cage motor and, upon relaxation of the excessive mechanical load, it will again lock into synchronous speed. It is also a motor that is amenable to experimentation. Indeed, empirical tactics seem the best path to optimizing its operational features.

An ordinary squirrel-cage rotor has additional symmetrically-spaced slots cut into it. The number of slots is equal to the number of poles. Soft-iron bars are then 'wound' into these slots. It is as if we make a magnetic

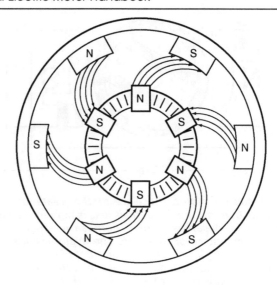

Fig. 3.11 *The reluctance motor in the act of pulling into synchronism. Clockwise rotation of the armature is assumed. Note that induced poles have developed in the embedded soft-iron elements.*

counterpart of a wound-rotor induction motor. The soft iron is magnetically different to the surrounding core material. In motor operation, as the speed approaches synchronism from the squirrel-cage motor action, the soft iron becomes magnetized sufficiently to form salient poles, enabling the motor to attain and lock into synchronism. This is shown happening in Fig. 3.11.

Single-phase fractional horsepower motors of this kind are useful for a wide variety of household appliances, where they overlap other types, such as repulsion motors and capacitor split-phase motors. The reluctance motor is convenient to manufacture because it utilizes mature technologies based on squirrel-cage rotors and conventional centrifugal switches. The modifications needed are easy and economical to accommodate. And, as with its first-cousin, the hysteresis synchronous motor, its efficiency can be reasonable. It is noteworthy that when the reluctance motor is running in its synchronous mode, there is virtually no current or attendant I^2R losses in the squirrel-cage. Also, the hysteresis loss of the soft-iron poles is very low.

The wound-rotor induction motor

A particularly interesting and useful AC motor is the wound-rotor induction motor. It is unfortunate that most are made only in large integral-horsepower sizes and are designed for industrial applications. One could easily devise unique applications for smaller versions of this motor in electronic systems. The main reason such motors haven't seen wide application in household appliances or for the mundane tasks of residential neighbourhoods is that they are polyphase (generally three-phase) machines. An awareness of the salient features of the wound-rotor induction motor is, however, bound to yield useful insights into the nature of other types of motors and to spark ideas for experimentation.

Because they operate from a polyphase power line, the wound-rotor induction motor sets up a rotating magnetic field in its stator and is self-starting. This contrasts to the various 'artificial' starting techniques needed for single-phase induction motors. The stator is essentially the same as might be used in a polyphase squirrel-cage motor, a polyphase synchronous motor, or a polyphase synchronous alternator. The unique element is the *rotor*. As implied, it is wound, rather than comprising single short-circuited conductors as in the squirrel-cage motors. The rotor resembles the drum-wound armature of a DC machine.

In three-phase wound rotor motors, there are three rotor windings connected to form a 'Y' configuration. The three free ends of the Y are brought out to slip-rings so that the rotor circuit can be completed through an external variable resistance-box comprising three Y-connected resistances. Such an arrangement is shown in Fig. 3.12.

With appropriate adjustment of the net rotor-resistance, the motor can be endowed with *low* starting-current, but *high* starting-torque. Once running, a wide speed range is selectable with the variable resistance. With the external rotor-resistance set to zero, the motor approaches the characteristics of a squirrel-cage motor, operating just below synchronous speed and with fairly good speed-regulation. Rotation reversal is readily accomplished by transposing two of the three line-leads. The speed-control possibilities of this motor are revealed in the performance-curves of Fig. 3.13.

It is commonplace to associate Selsyn systems with World War II radar technology. However, the discovery that wound-rotor induction motors were naturally adapted for such service was made even earlier. For example, the Panama Canal put this observation to practical use with regard to the operation of the locks. The operator had before him a miniature model of the canal locks. Movements of parts of the model were made to correspond to the movements of the parts of the actual canal locks via the synchronized rotor-positions of wound-rotor induction motors.

The connection diagram of two such motors is shown in Fig. 3.14. Such a pair of motors constitutes a Selsyn system. As would be expected for the canal installations, the machines are three-phase types. This contrasts with the small single-phase types used to synchronize the radar antennas and indicators in military systems. The operating principle remains the same regardless of phases. It is well to note that single-phase synchro transmitters and receivers do not require special starting techniques as do single-phase induction motors.

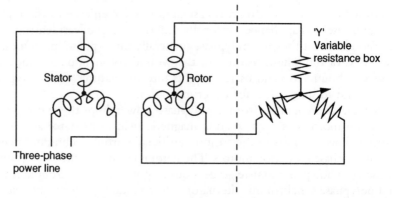

Fig. 3.12 *Control set-up for the wound-rotor induction motor. Slip-rings and brushes enable insertion of external resistance in the rotor circuit. This resistance can be adjusted to provide high starting torque together with low starting current. With the motor running, the resistance can be adjusted to provide wide-range speed control. Although not as efficient as the squirrel-cage motor, the operational flexibility of the wound-rotor type makes it the choice for many industrial applications.*

An interesting feature of Selsyn systems using large polyphase wound-rotor induction motors is that the receiver motor can exert considerable torque where big machinery is involved. In contrast, Selsyn systems in radar installations were usually called upon to provide positioning-accuracy foremost. This was attainable by relaxing torque requirements. In a sense, machines employed in Selsyn systems lose their identity as motors, for continuous unidirectional rotation is not the order of the day.

It is noteworthy that no single-phase wound-rotor induction motor has been marketed. It is conceivable that such a machine, with a shaded-pole stator (for starting) and with slip-rings to allow connection to an external variable resistance, might enable a usefully-wide speed range to be obtained for fan-type loads. Suggestive evidence of such operation was not lost to the author during his student days when a running three-phase wound-rotor induction motor accidentally became single-phase powered.

There is yet another interesting and practical use of the wound-rotor induction motor – its ability to operate as a *frequency changer*. Let the stator of such a machine be connected to a three-phase 60 Hz power line. For the moment, let the Y-connected rotor windings be open circuited, i.e., no load is connected to the slip-rings. Let the shaft of this motor be coupled to the shaft of another motor, say a DC shunt motor. The set-up is simple enough and it can be gleaned that we are about to use the wound-rotor machine as some kind of a generator or alternator.

Suppose the DC motor is energized and adjusted to turn at 300 r.p.m. Now, 300 r.p.m. is 1/6 of the synchronous speed of the wound-rotor machine's rotating magnetic field (assuming it is a four-pole motor). This

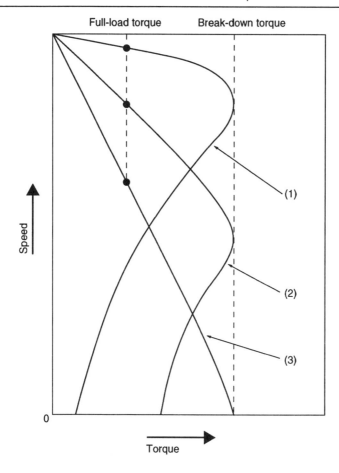

Fig. 3.13 *The unique speed characteristics of the wound-rotor induction motor. The curves correspond to three overall rotor-resistances as follows:*
(1) Lowest resistance.
(2) Medium resistance.
(3) Highest resistance. This is also a special *resistance, such that the overall rotor-resistance is equal to the reactance of the rotor. This yields the highest power factor for the rotor, together with* maximum starting torque.

implies that the slip of the rotor is 5/6 of the synchronous speed of its 60 Hz stator field. As a consequence, a voltage will be induced in the rotor windings having a frequency of 5/6 of 60 Hz, or 50 Hz. The voltage thus derived from the rotor will provide three-phase 50 Hz power to an appropriate three-phase load. In similar fashion, other frequency transformations can be made.

A natural refinement of this frequency-changing technique is to use a

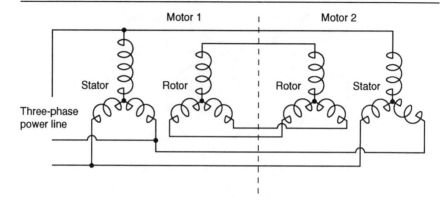

Fig. 3.14 *Two three-phase wound-rotor induction motors in a Selsyn relationship. The rotors follow one another so that their positions are synchronized. Either machine can be 'transmitter' or 'receiver'. Note rotor connections where like-terminals are connected. At mutual alignment of the rotors, their voltages are in opposition and no current flows in the rotor circuit. When there is departure of the rotor alignments, there is torque-producing current in the rotor circuit. This torque is always restorative, forcing re-alignment between the rotors. Thus, the rotors follow one another.*

synchronous motor in place of the above postulated DC motor. Then, by appropriate selection of the number of poles in each machine, various practical transformations of frequency can be precisely affected. For example, one could arrange to interconnect 50 Hz and 25 Hz systems. Similarly, a 400 Hz power system could be made available for testing the performance of aircraft motors.

The implication of this scheme should not be lost to the motor enthusiast. It clearly demonstrates that the frequency of the AC in the rotors of induction motors is determined by the slip, which is the difference between the mechanical speed of the rotor and the synchronous speed of the magnetic field in the stator. Thus, at a standstill the rotor frequency is that of the power line frequency. Conversely, at rated full-load speed, which may be, say 95% of synchronous speed, the rotor frequency will be very low – just 5% of the power-line frequency.

The double squirrel-cage induction motor

Some of the discussions of off-beat applications of the wound-rotor induction motor may have appeared remote from our everyday experiences with motors. However, it will be appreciated that some of the behaviourisms of the wound-rotor motor bear quite-relevantly on the principles underlying operation of a very popular motor that is extensively used in a wide variety of applications. This is the so-called *double squirrel-cage induction motor*. These

Practical aspects of AC motors 81

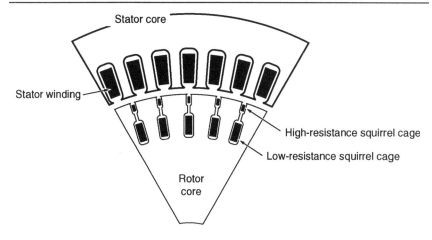

Fig. 3.15 *Basic constructional feature of the double squirrel-cage induction motor. The low-resistance cage also has* high inductive-reactance *because it is surrounded by a great deal of rotor core material. Conversely, the high resistance cage has relatively* low *reactance. During* starting, *the rotor frequency is high so the preponderance of torque-producing current flows in the high-resistance squirrel-cage. At higher speeds, rotor frequency decreases and induced current gradually diverts to the low-resistance cage. This favours desirable* running *characteristics for the motor. Noteworthy is the fact that no centrifugal switches are needed.*

motors' usefulness stems from the fact that they combine some of the salient features of both, the 'ordinary' squirrel-cage motor and the wound-rotor squirrel-cage motor. These are predominantly polyphase motors.

Specifically, the double squirrel-cage motor provides both high starting-torque *and* good speed-regulation with respect to load. To accomplish this, *two* squirrel-cage structures are carried by the rotor. Contrary to what one might expect, no centrifugal switch is used to transfer operation from one cage to the other. Yet, between standstill and rated speed, the transfer gradually and automatically does take place!

The two squirrel-cage structures are shown in Fig. 3.15. The inner cage features low resistance, but relatively high inductive reactance because it is of large cross-section and is immersed deeply in the rotor core-material. At standstill and at slow speeds when the rotor frequency is high, this cage will be largely ignored insofar as concerns induced current. Rather, most induced current will be in the outer cage which has high resistance, but quite low inductive-reactance. This is ideal for high starting-torque. (In the wound-rotor motor, this was achieved via an external resistance-box.)

As the motor gains speed, the rotor frequency decreases, enabling more and more induced current to be developed in the inner, low-resistance cage because its reactance no longer impedes such current. This too is good news, for good speed-regulation of the running motor requires a very low-resistance squirrel-cage. At operating speed, the high-resistance cage can be said to be ignored as a participant in contributing to motor torque. This is because of the much-higher current (and better power-factor) now carried by the low-resistance squirrel-cage.

Speed control of AC motors

AC induction motors and synchronous motors are surprisingly amenable to wide-range speed control if the *frequency* of the AC source can be varied. This was always obvious, but was never very practical or cost-effective until solid-state inverters became available. Even though an induction or synchronous motor may have been designed and rated for 50/60 Hz, it will likely yield very nearly an equally efficient service over an extensive frequency range, say from 30 to several-hundred Hertz. The motor speed will directly follow the AC frequency. For example a synchronous motor rated at 1800 r.p.m. at 60 Hz will run at 900 r.p.m. at 30 Hz and at 3600 r.p.m. at 120 Hz. There are, however, a few caveats one must abide by, so let's first deal with polyphase machines.

What is sometimes forgotten is that provision must be made to either automatically or manually vary the applied voltage in direct proportion to the frequency. Thus, if the rated operating voltage of our 60 Hz synchronous motor is 120 V, it must be impressed with 60 V at 30 Hz and 240 V at 120 Hz. This applies in the same way to induction motors.

So much for polyphase AC motors. Single-phase induction and synchronous motors can be speed controlled in much the same way, but with the proviso that something needs to be done about the centrifugal switch in the split-phase types. Since these switches are mechanically designed to open at about 80% of 60 Hz synchronous speed, they can no longer provide their intended function at lower or higher motor speeds. This will result in an overheated motor due to retention of the starting winding in the circuit. A neat practical solution is to use a permanent-capacitor induction motor inasmuch as this single-phase machine has no centrifugal switch to begin with.

Another solution will entail some experimentation, but is capable of good results. It involves the substitution of an electronic cut-out switch in place of the centrifugal switch. Incidentally, the centrifugal switch need not be physically removed, but only electrically-disconnected.

The consequent-pole AC motor

For decades, physicists have been searching for evidence of a theoretical entity known as the *monopole*. As suggested by the name, this is supposed to be a magnetic pole independent of any linkage to an opposite pole. It so happens that monopoles are used in electric motors but no demand is made

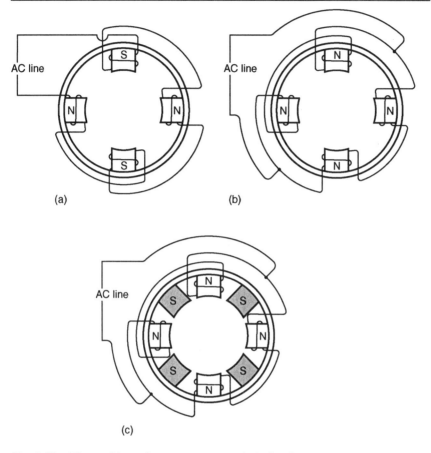

Fig. 3.16 *The making of a consequent-pole induction motor.*
(a) Original connections on stator of a single-phase four-pole squirrel-cage motor. Note alternate poles.
(b) Transposed connections to produce four 'monopoles'. However, such isolated poles cannot exist as such; rather, the consequent *poles depicted in (c) are produced.*
(c) The practical polar-pattern as a result of the modification made in (b). Note that there are now eight *magnetic poles rather than the original four. Accordingly, the motor now will run at* one-half *of its original speed.*

for their isolated existence. What is being referred to here is the so-called *consequent pole* AC motor in which the major poles, instead of alternating in polarity, are all of the same polarity, either N or S. In other words, one can rewire the pole connection in the stator of a single-phase induction motor, making transpositions causing all poles to be the *same*. As will be seen, these 'monopoles' unfortunately do not satisfy the physicist's quest for isolated magnetic poles.

Indeed, the main poles now produce opposite poles between them so that the actual polar pattern again consists of a circular sequence of *alternate* poles. This is shown in Fig. 3.16. We now have *twice* the number of pole-pairs as originally. The new in-between poles are known as *consequent* poles. The motor will now run at *half* its original speed, just as if it had been constructed with the same number of true poles. This technique is widely used for changing the speed of fans. Sometimes it is combined with changes in the number of active poles in order to provide even more speed selections. Significantly, the squirrel-cage doesn't care *how* the pole-pairs are produced – its speed is geared to the total number of pole-pairs whether real or subsequent. Do not confuse this technique with the *interpoles* use to combat armature reaction in DC motors. Although the interpoles are inserted between the main poles, they do not develop magnetic fields oriented properly to act as main fields. They are wired in series with the armature so that stronger armature reaction is automatically more strongly opposed. (Sometimes, interpoles are *inductively* coupled to the AC in the armature of DC machines.)

Returning to the AC induction motor; although the consequent poles may not be physically obvious, they are very real magnetically. Interesting experiments could be carried out in applying the consequent-pole technique to synchronous motors.

Speed selection by pole modification

AC induction and synchronous motors always have their stators wound for an *even* number of poles. This is because of the nature of magnetic poles; they always occur in dual arrangements of a north and a south pole. As shown in our previous discussion of the consequent-pole motor, an attempt to produce an isolated monopole fails because an alternate pole is spontaneously induced to form the required N–S pair. Of course, this fact was used to advantage in the consequent-pole motor where the additionally-formed poles brought about *lower* motor speed.

It fortuitously happens that *higher* motor speeds can be attained via a simple method of *reducing* the effective number of poles. First, however, it would be well to keep in mind that the actual number of operative poles provided by the stator windings depends upon the number of phases. Thus, the six separate windings on the stator of a three-phase motor tells us that the speed will be that of a two-pole machine. Similarly, eight *apparent* poles would correspond to four-pole operation in a two-phase motor. (Note, that the six-pole stator is inappropriate for a two-phase motor because motors must have *even* numbers of functional poles.)

Life is simpler for single-phase motors. Consider a four-pole stator of a single-phase induction motor. (Such a motor will, of course, have some 'gimmick' for self-starting, such as an auxiliary starting-winding and a

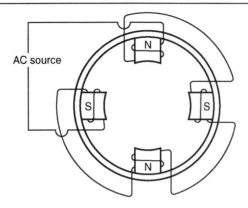

Fig. 3.17 *Speed doubling by reconnecting stator field windings. In this scheme, an original four-pole stator is made to behave as a two-pole winding. Thus, a four-pole induction or synchronous motor that originally ran at or near 1800 r.p.m. from a 60 Hz line, would now become a two-pole machine and its operating speed would now be at or near 3600 r.p.m. (At first glance, the unorthodox sequence of the instantaneous poles in the above diagram could be easily overlooked.)*

centrifugal switch.) In contrast to stator winding reconnections we made in the consequent-pole motor, we can this time reconnect the windings so that there are *two* large-poles in place of the original four smaller-ones. That is, a pole and its neighbour-pole will both be N-poles. And the oppositely-facing pair of poles will both be S-poles. This, of course is just a matter of reversing the current direction in alternate poles. Strange as it may seem, we now have a two-pole motor. The practical manifestation of this will be operation at *twice* the speed of the original four-pole design. This technique can be extended for both single and polyphase induction and synchronous motors. See Fig. 3.17.

Generally, this speed changing method is used for fan motors. The torque demands of such an application are not severe. In practice, negligible side-effects assert themselves from such speed control of the 'constant-speed' AC motors.

Another speed-selection technique

A more direct method of altering the number of stator poles is shown in Fig. 3.18. Here, by means of *multiple windings*, it is possible to select the number of active poles and thereby change the speed of the motor. This is a straightforward approach, one that is likely to be favoured by the manufacturer. When the switch connects the AC power-line to contacts 1 and 2, all

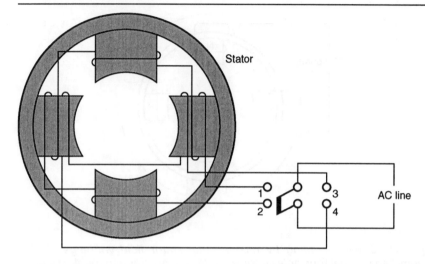

Fig. 3.18 *Alternative technique for speed selection in AC motors. Multiple windings enable the selection of the number of operative poles in the stator. In the above diagram, switch position 1–2 makes use of all four poles. Twice the speed results with the switch in position 3–4 because only two poles are then used.*

four poles are active and the motor is in its 'slow' mode. When the switch connects the AC power line to contacts 3 and 4, only the two horizontal poles are active. The motor then operates at double speed; it is now in its 'fast' mode.

The methods described for either increasing or decreasing the number of active poles have provided effective speed selection for fan applications where any degradation of efficiency, power-factor or torque is of little practical consequence. A more modern speed-control technique is to provide a variable frequency AC source. (At the same time, the applied voltage should be changed in inverse relationship to the frequency.) If the adjustable speed range is not too wide, adverse effects can be negligible and near-rated loads can be accommodated. One thing to keep in mind, regardless of how the departure from 'rated' speed is brought about, is that the self-ventilation of a motor falls off rapidly as its speed is reduced.

A problem with three-phase induction motors

Since the inception of industrial electrification, the three-phase induction motor has been the workhorse of manufacturing. To anyone basically acquainted with the nature of electric motors, the underlying reasons are quite obvious. This motor format is self-starting and runs at nearly constant speed over a wide load range. It is easily reversible and it may be 'plugged' to bring it to a quick stop. It is very efficient and its power factor can be reasonably good. Best of all, it is simple and very reliable. The squirrel-cage types have no moving contacts or brushes so that maintenance requirements are exceptionally low. As if all this is not enough, modern solid-state inverter technology enables wide-range speed control by varying the applied frequency.

Notwithstanding these manifold features, a certain operational defect has long been known to happen occasionally. Suppose that one of these motors has been started from its three-phase power line and is driving its rated load. For a variety of reasons, it can happen that *one* of the three-phase lines opens, allowing electrical power to be supplied to the motor via just two lines. Such an event thereby supplies *single-phase power* to the motor. As might be anticipated, the motor's efficiency, speed regulation and power factor all suffer and the torque may be rough and there may be some audible 'protest' noise.

Surprisingly, however, in many instances there may be little that is immediately obvious and the motor may obediently do justice to its load. We are reminded here that single-phase induction motors, *once started*, are capable of good performance. It is true however that for the same machine weight, the three-phase format can develop about 60% more shaft horsepower than single-phase induction motors. From this, we can infer that our three-phase motor is likely to overheat if driven too long by single-phase power. If this situation is suspected, it can quickly be investigated by applying a clamp-on ammeter to the three leads that bring three-phase power into the stator windings of the motor – if conditions are normal, these three readings will be substantially the same.

The universal motor

It is commonly pointed out in texts that the DC series motor will continue to function well and rotate in the same direction if the polarity of the DC supply is reversed. This makes sense, for the interacting magnetic fields which produce the mechanical torque *both* undergo flux direction reversals. There is therefore no change in the direction of rotation of the armature. (One must reverse only one of the interacting fields in order to reverse rotation.) It is only natural to ponder the operation of such a motor on alternating current. These matters are probably 'old-hat' to many readers. Nonetheless, certain difficulties are often encountered in practice wherein nebulous notions about universal motors tend to be responsible for the problems.

A true DC series motor, although demonstratively operative on AC, is unlikely to function optimally. Its efficiency can be appreciably lowered because of eddy-current and hysteresis losses. Also, the effects of armature reaction may no longer be negligible; this will be evidenced by inordinate sparking of the brushes. Some hobbyist objectives may still be met, but good engineering practice will have to either scale down the load or duty-cycle requirements, or resort to special circuit or operational techniques.

On the other hand, an AC series motor is more likely to deliver acceptable performance when operated from a DC source. Replication of behaviour is easier to obtain at high than at low speeds. To combat the effects of armature reaction, some of these AC motors have inductively-coupled compensation windings in their pole faces. Surprisingly, these can continue to yield good commutation with the motor running from a DC supply. This is because there is alternating current in the armatures of all DC machines.

Although not a hard and fast rule, AC series motors in which the compensating windings are *conductively-connected* in series with the armature tend to be better bets as true universal types. Small series-type motors with no compensation windings can behave well on both DC and AC if otherwise designed for such universal application. It makes good sense to use a motor in which 'universal' appears on the nameplate.

Rotation reversal in AC motors

As we have seen, the direction of rotation of the classic DC motors can be abruptly reversed by interchanging a couple of external or internal leads. The nice thing about this is that this can take place while the motor is delivering rated horsepower output and speed. Unfortunately this is not always true with AC motors.

Because polyphase AC motors generally develop reasonably high starting torque, and because they tend to have definitely-defined rotating fields, these motors may also be 'plugged'. That is, they may be fed reversing information while carrying full load at rated speed. Included here are all types of two and three-phase induction motors and synchronous motors.

With single phase induction motors, some care has to be observed. In the popular resistance-start split phase motor or those incorporating a high-resistance starting winding, reversal by plugging is generally not practical – the motor can just continue running in the same direction. Note that these motors have relatively low starting torques and that their rotating fields are more elliptical than circular. (Much of the energy invested in the stator produces a pulsating, rather than a rotating field.) All is not lost, however. If either the main winding or the starting winding is reversed, the motor will then start from *standstill* and rotate in the opposite direction. All this likewise applies to the single-phase synchronous motor employing a high resistance or a high resistance starting winding for phase-splitting.

Many single-phase induction and synchronous motors utilizing *capacitors* for phase-splitting can be successfully reversed while running at rated speed and with rated load. Such motors develop a reasonably good two-phase rotating field. The motors, themselves, are basically two-phase machines.

Reversal by plugging may run into difficulties, however, if there is excessive inertia associated with the load.

The fractional-horsepower universal motors commonly encountered require a bit of minor surgery to reverse their direction of rotation. These are generally uncompensated types and it is only necessary to transpose the series-field connections. Sometimes the brush–commutator surfaces do not initially mate well on reversed rotation; this generally-remedies itself after a few hours of operation where it will be observed that random and erratic sparking gradually diminishes. Some universal motors, such as those that have been used in garage-door systems, conveniently have multiple leads so that the direction of rotation is switchable.

There are several versions of the repulsion motor. Basically, this AC motor can be viewed as a series or universal motor in which the series field, rather than being conductively-connected to the armature, is inductively coupled to it. The armature is short-circuited so that a heavy current can be induced in it. Because of this heavy current, the repulsion motor develops high starting torque, somewhat similar to a series or universal motor. However, in order for the repulsion torque to manifest itself, the shorted-brush assembly must be rotated something on the order of 20° from the neutral magnetic axis. See Fig. 3.19.

Herein lies the clue to reversal of the direction of rotation – a *mechanical* rather than an electrical modification is called for. Specifically, if the brush axis is moved forward (to the right) of its geometrical or neutral axis, clockwise rotation will ensue. Moving the brush axis backward from its neutral position will cause counter-clockwise rotation. For clarity, one can simply say that rotation will take place in the direction of brush-axis departure from the neutral axis or plane.

A more elegant motor utilizing the repulsion principle for high starting torque is the repulsion induction motor which dispenses with centrifugal mechanisms, yet combines the best features of both repulsion and induction motors. Such a motor has a double-function rotor. The outer periphery of the rotor contains a conventional drum winding such as is common in DC motors. Associated with this winding is a commutator and a short-circuited brush assembly as is common in repulsion motors. There is an angular displacement between the brush axis and the neutral axis. (This displacement, which is necessary for torque development, may be brought about by either physical or electrical means.) The unique feature of this rotor is that it also contains a deeply embedded squirrel-cage structure. Because of the high inductive reactance of this squirrel-cage, it has only a small current in it at standstill and at slow speeds. At higher speeds, because the slip frequency is low, higher current is induced in the squirrel-cage conductors and motor characteristics become more like those of the induction motor. The stator of these machines is the same as those used with induction motors.

The rules of rotation reversal described for the other types of repulsion

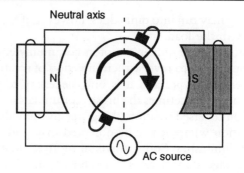

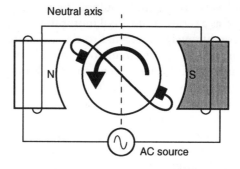

Fig. 3.19 *Directional rotation of the basic repulsion motor. Angular departure between the neutral axis and the brush axis is required for the development of torque. The direction of rotation depends upon whether this angle is positive or negative.*

Although practical repulsion motors tend to be multi-polar with pairs of brushes for each pair of poles, these simplified sketches illustrate the principle involved in reversing the direction of rotation.

motors also apply to the repulsion induction motor. It can be appreciated that many permutations are possible with the various types of repulsion motors. First on the agenda, one must determine the design features of the motor at hand. Then, from a basic understanding of the operating principles, the direction of rotation can usually be achieved whether or not such operational flexibility was intended by the manufacturer. This is a valid approach for most experimental purposes, but one should not be surprised to find the maker's warranty voided.

Instead of physical displacement of the brush axis, some repulsion motors make use of a more sophisticated technique to develop torque, although the net result is the same. In these motors, the brush axis is not adjustable, but is coincident with the neutral axis. Each polar winding on the stator consists of *two* sections connected in series. The two winding-sections are angularly displaced from one another on the stator. This produces a purposeful twisting-distortion of the main field flux so that the neutral axis is no longer symmetrically located between poles. In other words, an angular displacement between the neutral axis and the brush axis is achieved. This is the *same* situation that prevails in motor designs incorporating adjustable brushes.

With some minor surgery, these repulsion motors can also be reversed. This is done by transposing the series-connections already referred to. Such intervention will bend the field flux in the opposite way, bringing about the desired reversal of rotation. Such a modification is electrically-equivalent to physically moving adjustable brushes in the opposite direction.

In real life, one is more likely to become involved with combination-type, rather than the 'pure' repulsion motors we have dealt with. Such motors *start* as repulsion motors, but change over to *induction* motors as they attain speed. The way in which they accomplish this operating transition gives rise to the different types.

In one type, known as the repulsion-start induction-run motor, a centrifugally operated mechanism shorts *all* commutator segments at about 80% of synchronous speed. This makes the rotor behave as a squirrel-cage and the operation then commences as a squirrel-cage induction motor. The best of both worlds is thereby attained – high starting torque, together with good speed regulation. In some models, the brushes are also automatically lifted from the commutator in order to reduce maintenance.

Non-sinusoidal waveforms applied to AC motors

A practical problem often encountered concerns the operation of various AC induction and synchronous motors from solid-state inverters. For example, the inverter may be used to enable operation of various electrical appliances from a 12 V vehicular battery. The nameplate on the electric motor will, of course, state the operating voltage, the assumption being that this voltage is of sinusoidal wave-shape. The AC output of the solid-state inverter on the other hand is likely to be a square wave. The question arises regarding the magnitude of this square wave that should be applied to the motor in order to obtain equivalent performance to that obtained from 'normal' sine-wave operation.

The superficial simplicity of the situation is deceptive and common sense all too easily produces the wrong answers, or the right answer for the wrong reason. One's knee-jerk analysis might suggest that there is more energy-content in a square wave than in a sine wave of the same r.m.s. value and that the motor would therefore develop greater torque. This could lead to the conclusion that a *lower* r.m.s. square wave would suffice to make the motor operate with rated performance. Experiment could conceivably show this to be so, but the confirmation would nevertheless be based upon erroneous reasoning.

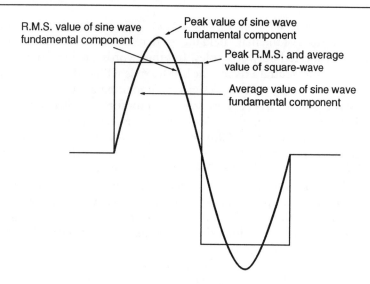

Fig. 3.20 *A symmetrical square wave and its fundamental frequency component. The waves are not what one sees on ordinary oscilloscopic displays, but derive from the mathematical treatment of Fourier analysis. Although the peak of the sine wave exceeds the peak of the square wave, both the r.m.s. and average values of the sine wave (the fundamental frequency) are lower than the square wave values.*

We know that the square wave drive is comprised of a fundamental frequency and many odd-harmonics of the fundamental. It is the fundamental component that does the actual work in causing motor action. The harmonics contribute to the temperature rise through the agencies of eddy-currents and magnetic hysteresis. Worse, they interfere with the development of the desired electromagnetic torque in the motor. In other words the 'extra' energy content of the square wave is not at all useful for motor operation. We come now to a paradox generally overlooked in the intuitional appraisal of this situation.

Figure 3.20 reveals some interesting features of a symmetrical square wave. For the sake of simplicity, only the *first* harmonic, that is, the fundamental frequency, is also drawn in. We have what appears to be an awkward situation in which the peak of the fundamental exceeds the amplitude of the square wave. (If a great number of the higher harmonics were also drawn in, the ultimate result would be the square wave itself and there would be no challenge to our concept of reality.) As pointed out, our justification for dealing with only the first harmonic is that it is the major contributor to torque in most AC motors.

It will be noted that even though the peak value of the first harmonic (fundamental) sine wave exceeds that of the square wave, its r.m.s. value is *less* than that of the square wave. Quantitatively, the r.m.s. value of the

Practical aspects of AC motors 93

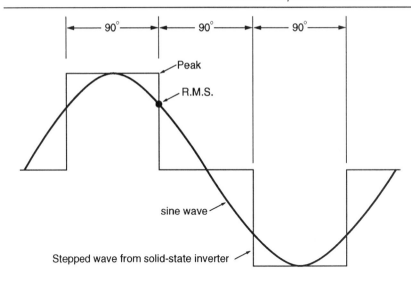

Fig. 3.21 *Stepped wave claimed to be especially suitable for driving AC motors. With the 90° step, this waveform has the following features:*
(1) The troublesome third harmonic of square waves is low.
(2) It resembles a sine wave in that it has the same r.m.s. and peak levels.

fundamental frequency sine wave is only 89.7% that of the square wave. Or stating the situation the other way around, the r.m.s. value of the square wave is 1.11% greater than the r.m.s. value of the fundamental. On this basis, the r.m.s. value of the square wave delivered by the solid-state inverter should be 1.11 times the *rated* operating voltage on the name-plate of the motor. Thus, a motor rated for operation at 120 volts AC should receive 1.11×120 or about 133 volts.

If this is attempted, however, it may be found that the motor develops excessive temperature rise from the eddy-currents and hysteresis produced by the third and higher harmonics. On the other hand, some motors will present sufficient inductance to attenuate the harmonics to the extent that no bad effects are noticeable. Even though the applied voltage wave may be very nearly square, the motor inductance tends to cause the current wave to be trapezoidal, or nearly sinusoidal in some instances. It is the *current wave* that develops the *torque* in the motor. It is easily seen that one must resort to experimentation and that different results can be anticipated for different motors and motor loads. Keep in mind that stepping motors harness the *extra* energy in square waves, but induction and synchronous motors strongly prefer sine wave drive.

Let us now look at another non-sinusoidal wave available from some solid-state inverters. This is the stepped wave depicted in Fig. 3.21. We shall be specifically interested in the unique pattern wherein the step is equal to

90 *electrical*-degrees. A Fourier analysis of this wave would reveal the surprizing fact that the third harmonic is quite low. It is the third harmonic of the simpler square wave that causes most of the troubles when impressed upon motors. At best, the third harmonic is non-usable energy, contributing nothing to torque production in the motor.

For practical purposes, it turns out that a motor can be almost as 'happy' when driven by a voltage source with this stepped waveform as with a sine wave. Indeed, one could postulate that the motor simply treats such a wave as a sinusoid in the rough. It will be observed that this waveform has both peak and r.m.s. values identical to the ideal sine wave. Indeed, with this applied voltage waveform, the torque-producing current wave is likely, because of motor inductance, to actually resemble a fairly-good sine wave. The bottom line here is that the stepped wave is treated *as if* it were a sine wave in complying with the ratings inscribed on the nameplate of the motor.

It can be easily construed that contradictory logic is employed in dealing with the square wave and the stepped wave. This stems from the relatively-high energy content of the square wave, but one must keep in mind that most of this 'excess' energy resides in the third and higher harmonics and is *not* usable to the induction or synchronous motor. Universal motors generally have sufficient inductance to smooth the current wave. Nonetheless, the square voltage-wave can cause eddy current and hysteresis heating, and can produce abnormal brush-sparking.

Regardless of motor type, it may be found that a crude low-pass filter made up from a few turns of heavy wire on a ferrite core and a large capacitor can alleviate rough running problems and temperature rise from the harmonic energy of square waves.

Power input and power factor in three-phase induction motors

Because of the widespread use of three-phase induction motors in industry and manufacturing, it is often desirable to know the power consumption and the operating power factor of these machines. A handy technique for determining both is by appropriate use of two identical wattmeters. The scheme is quite straightforward, but discipline is required in the interpretation and deployment of the recorded data. The set-up for such data acquisition is shown in Fig. 3.22. The partial-pictorial aspect of Fig. 3.22 is intended to emphasize the importance of exactly emulating the physical layout. Once this is accomplished, the following conditions should be kept in mind:

(1) The three-phase AC line should provide balanced voltages. There is a balanced load requirement too, but this is automatically met by the motor.

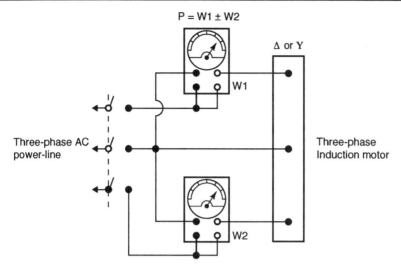

Fig. 3.22 The two-wattmeter method of three-phase power and power factor measurement. This is a convenient technique for evaluating the performance of three-phase induction motors. Power is simply the algebraic sum of the power readings. Power factor is then calculable from these readings. The same results will be obtained regardless of which meter is labelled W1 or W2.

(2) It makes no difference whether the stator windings of the motor comprise a delta or 'Y' configuration.
(3) It makes no difference which wattmeter is W1 or W2. The *same* results will be obtained.
(4) In the equations for power and for power-factor, W1 and W2 combine *algebraically*. This is because either of these quantities can be *negative*. Indeed, for load situations where the power factor is less than 50%, *one* wattmeter will yield a negative reading.
(5) Here's how to identify a negative reading: whichever wattmeter requires reversal of its current coil in order to obtain an upscale deflection has been yielding a negative reading. (Some wattmeters incorporate a switch for conveniently accomplishing the reversal.)
(6) Here are some guideposts for getting the 'feel' of these measurements:
   (a) For lagging power factors less than 50%, W1 will be negative.
   (b) For leading power factors less than 50%, W2 will be negative.
   (c) For lagging *and* leading power factors greater than 50%, both wattmeter readings will be positive.
   (d) At 50% leading power factor, W2 will read zero.
   (e) At 50% lagging power factor, W1 will read zero.
   (f) At unity (100%) power factor equal positive readings will be displayed by W1 and W2.

(g) A loaded induction motor will generally operate at a greater than 50% lagging power factor. *Positive* readings from W1 and W2 will be obtained.

(h) A very-lightly loaded, or unloaded induction motor can operate at less than 50% lagging power factor. Watch for a negative W1 reading.

In actual practice, the measurement procedure turns out to be quite simple, as will be demonstrated in the following example.

Two identical wattmeters are connected to a three-phase induction motor as shown in Fig. 3.22. The readings are 2000 watts (W) and 800 W. Because the motor is being operated at near full-load, it is known that both of these readings are *positive*. It is desired to know the *power-input* to the motor and the corresponding *power-factor*.

(1) The algebraic and arithmetical sum of the readings are the same because they are *both* positive. Therefore, Power Input = W1 + W2 = 2000 + 800 = *2800 W*
(Note the arbitrary designation of wattmeters, W1 and W2.
The reverse situation would be equally satisfactory.)

(2) For power factor,

$$\tan \theta = \sqrt{3}\left(\frac{W1-W2}{W1+W2}\right) = \sqrt{3}\left(\frac{2000-800}{2000+800}\right) = 1.73\left(\frac{1200}{2800}\right) = 0.7415$$

(3) The power factor is the cosine of the angle corresponding to the value of the tangent, 0.7415. This is looked up in a table of trigonometric values. It is found that the angle θ is about 36°33' and the *corresponding* cosine is approximately 0.803. By definition, *power factor = cosine* θ = 0.803
NOTE: At no-load, one of the wattmeter readings, because of the lower than 0.50 power factor probably involved, could have been *negative*. However, the *same* procedure, with due regard to the algebraic signs, would apply.

Unusual motor behaviour

Both by intent and by way of malperformance, motors sometimes behave in unusual ways. This is particularly true with AC motors. In a way, it is not altogether surprising considering the many things that occur more or less at the same time. This includes hysteresis, eddy-currents, armature-reaction, non-linear magnetics, skin-effects, and harmonic currents. An example of a type of malperformance is the tendency of some synchronous motors to lock at a sub-synchronous speed during start-up.

Interestingly, small motors are marketed that are deliberately designed for sub-synchronous speed operation. These can take the form of single-phase

Practical aspects of AC motors 97

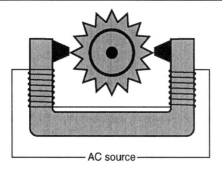

Fig. 3.23 *A simple type of sub-synchronous motor. The basic feature of a variety of such motors is the high excess of rotor over stator poles. The illustrated structure has 16 salient rotor-poles and two stator-poles. It is not self-starting and must be initially spun by hand. At 60 Hz, its synchronous speed, governed by the rotor teeth, is 450 r.p.m. If excessively loaded, it can 'shift' downward to one-half, or one-third of the 'fundamental' synchronous speed.*

types similar to the shaded-pole motors previously described. However, the sub-synchronous motor has a gear-wheel shaped iron rotor with a large number of teeth. The motor attains a synchronous speed dictated by these teeth, or salient poles, even though excited by a two-pole stator. For example, at 60 Hz such a motor with sixteen teeth will rotate at 450 rpm. If, however, such a motor is loaded heavily enough to exceed its pull-out torque, it will obligingly shift downward to one-half, or one-third of its basic synchronous speed and will maintain constant speed at the lower rpm values. Nor is that all; this electrical 'gear-shifting' will be accompanied by development of greater torque. A basic subsynchronous motor is shown in Fig. 3.23.

Ordinarily, for an induction motor to perform as an induction generator, connection to an AC power line is required. Then, generator action takes place if instead of supplying torque, the motor is *driven* above its normal synchronous speed. In developing countries where no nearby AC line exists, a novel scheme can be used to convert the hydro-energy from a stream into electrical energy. It so happens that three-phase induction motors are widely available at relatively low cost. These motors can be deployed as stand-alone induction generators of single-phase power. In such use, the generators are caused to be self-exciting by connecting appropriate capacitors across the phases. See Chapter 5 for the calculation of the capacitors. Residual magnetism accounts for the initial build-up in these motors-turned-generators.

Engineering handbooks sometimes state that the result of overloading a synchronous motor so that its pull-out torque is exceeded will be a slow-down to standstill. This is essentially true for the classic three-phase synchronous motor with damper windings in its pole-faces. It is easy enough to imagine, however, that *if* the induction motor action of the damper windings was great enough, the motor would simply drop its synchronous speed but would continue to run at a slightly-slower speed in the manner of a squirrel-cage induction motor.

It so happens that certain synchronous motors that do not use DC-excited rotors, but rely on either reluctance or hysteresis for the formation of rotor poles, *will* continue operation as induction motors when excessively loaded. Moreover, these motors will again accelerate to synchronous speed when their loads are relaxed. In one of these motors, known as the synchronous induction motor, there is a full squirrel-cage winding on the rotor. At the same time, the steel rotor is so-shaped as to enable 'salient' poles to form in the presence of the field from the stator. Single-phase versions of these motors, often called reluctance motors, can behave in an essentially similar way with regard to excessive loading. (For starting, however, the single-phase type make use of starting and running windings on their stators, and incorporate centrifugal switches.)

Despite the diverse techniques used in the various types of synchronous motors they are all capable of maintaining constant average speed when operated from a constant-frequency AC source and when carrying loads within their ratings. A number of sophisticated instrumentation systems are based upon the use of a synchronous motor driven from an AC source derived from a very stable crystal-oscillator. In such systems, the synchronous motor is protected from both voltage and mechanical transients so that there is no cause for even a momentary deviation from synchronous speed.

4 Practical projects

Experimental aspects of electric motors

In dealing with the practical aspects of electric motors, it is only natural to focus on some things that can be done by experimenters, innovators, and the general run of hand-minded practitioners. In so doing, the use of common hand-tools, rather than exotic machine-shop equipment will be emphasized. Many electrical and mechanical modifications can be made that fall far from the realm of major surgery. Various operational characteristics can be obtained by simply requesting an appropriate IC control module from the semiconductor firms. Here one enjoys the luxury of 'tailoring' performance electronically, sometimes without too much regard for the motor's 'natural' behaviour.

Doing things with motors other than simply complying with nameplate constraints can be both instructional and fun. Although it is convenient to deal with relatively-large motors in the text, it is obvious that small counterparts are widely available at low cost and with greatly reduced possibilities of disrupting electrical service, damaging feeder lines or exposing the experimenter to electrical or mechanical hazards. Moreover, it often makes good sense to first alter a small machine, record all relevant data, then scale up the level of the modifications if a larger machine is desired.

Considerations in starting electric motors

Motor starters momentarily or temporarily reduce the voltage and/or the current drawn from the power line when a motor is first energized. In motors over 5 HP, and certainly in multi-horsepower motors, connecting the motor immediately across the line results in a high inrush current that can damage the motor. Additionally, one must safeguard fuses and circuit breakers, as well as spare incandescent lights from excessive dimming. At start-up, DC motors generate no counter-e.m.f. and are exposed to many times rated operating-current because of their low armature resistance. AC induction and synchronous motors behave as transformers with short-circuited secondaries at the moment of power application. With all types of motors, start-up tends to be accompanied by a higher level than the operating current while accelerating to ultimate running speed.

Motors of say, one-half or smaller fractions of one horsepower ordinarily can be thrown directly across the line. This includes most household appliances and those used in electronic technology. Motor starters used with multi-horsepower motors comprise an art in itself. Various means are used to initially reduce current consumption, such as series resistance, series reactance, transformers, auto-transformers, and taps on the motor-winding themselves. Starting fully-loaded motors imposes additional demands on the starting procedure because the motors then require more time to attain their rated running speed, thus prolonging their near-short-circuit behaviour.

When considering the implementation of motor starters one must decide whether the start-up sequencing will be done manually or in a nearly-automatic manner. It is important to make the procedure 'idiot-proof'. In large installations, bad timing or confused judgement can produce violent disturbances.

Although the basic current-limiting function underlies all motor starters, different types of motors require unique switching and sequencing arrangements. Considerable tailoring of circuit parameters must be applied to particular installations, and because of a motor's torque requirements, some compromise is usually needed with regard to allowable current reduction. Utilities often impose mandatory rules regarding the incremental sequencing of current steps during start-up. In the starter shown in Fig. 4.1, the rheostat has negligible effect on field current. Note the protection against field current and line-voltage interruption.

Starting a large synchronous motor illustrates a special procedure as these motors are started as induction motors via the action of damper windings inserted in the pole-faces. These windings are actually short-circuited bars and behave as do the squirrel-cage structures in conventional squirrel-cage induction motors. (Neither single-phase nor polyphase synchronous motors are self-starting without the damper windings, also called amortisseur windings.) In the starting procedure of these motors, the field winding is sometimes shorted, but in any event, no DC field excitation is applied until the rotor attains near-synchronous speed. Application of field current at that time, causes the final incremental jump of the rotor to synchronous speed.

This starting procedure works well if the synchronous motor is lightly loaded. If the full load is in place, it may be found that there is insufficient starting-torque and in such instances a higher resistance damper-winding is necessary. Sometimes multiple turns are used so as to emulate the wound-rotor induction motor which gains its high starting torque via external resistance and of course, slip-rings must be added for such a provision.

It is obvious that a proper installation of large motors requires special attention. Fortunately, the starting requirements of the motors that the readers of this book are likely to be involved with can usually be satisfied

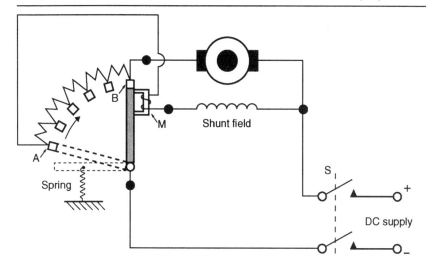

Fig. 4.1 *A typical starter for a DC shunt motor. After closing the main switch, S, the operator moves the rheostat arm from its spring-held rest position to the first (highest) resistance-contact, A. As the motor starts and gradually attains speed, the rheostat arm is slowly moved through the progressively-lower resistance contacts, finally culminating at the zero-resistance contact, B. At this position, with the motor at full-speed, the rheostat arm is held in place by an electromagnet, M. Lack of field current returns the arm to rest. Interruption or appreciable reduction of the line-voltage also releases the hold on the rheostat arm and its spring returns it to the rest position.*

with simple current-limiting and switching arrangements. Indeed, as has been mentioned, these fractional-horsepower motors are generally thrown across the line with minimal disturbance.

High starting-torque from a small capacitor

A variety of capacitor split-phase induction motors are available for operation from a single-phase AC line. On the one hand, the capacitor is an excellent element for bringing about the required quadrature phase-displacement. In other words, the rotating field thereby produced closely simulates that of a true two-phase system. There have been, however, some practical difficulties with this scheme of endowing the motor with starting torque.

Because the required capacitance tends to be quite large, especially for high starting-torque, electrolytic-type capacitors are necessary for such motors above 1/10 horsepower, or so. Electrolytic capacitors feature a high capacitance-to-volume ratio and their cost can be reasonable. However, being electro-chemical components, they are subject to ageing and to temperature effects. Deterioration also occurs due to the necessity of carry-

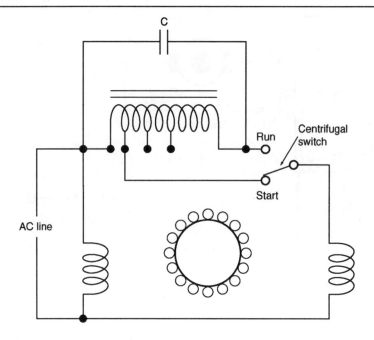

Fig. 4.2 *High starting torque by means of capacity 'magnification'. One relatively-small capacitor provides both high starting torque and efficient running characteristics. The motor at startup 'sees' the capacitance multiplied by the square of the auto-transformer's voltage step-up ratio. At about 75% of synchronous speed, the centrifugal switch alters the capacitor circuit, placing the motor in its* Run *mode. (What appears to be a possible* LC *resonant 'tank', actually has no practical effect.)*

ing high currents, and from being subjected to irregular operation-cycles. In many instances, the electrolytic capacitor becomes a high maintenance item.

A neat scheme for 'fooling' the motor with a low-capacitance oil-type capacitor is shown in Fig. 4.2. Here, an auto-transformer is used to reflect a much higher capacitance to the motor than exists physically. This is based upon the principle that a transformer reflects impedance from a secondary to a primary winding in direct proportion to the square of the secondary/primary voltage ratio. When deployed in motor circuits, a 7:1 step-up ratio is commonly used and this gives rise to a nearly 50:1 multiplication of the capacitance.

It is true that the capacitor, although of reasonably low capacitance, must have a high-voltage rating, which in itself is costly. Overall, however, the scheme is economically justified because the oil-type capacitor tends to be a low maintenance item and its characteristics remain relatively stable with respect to age and temperature.

In Fig. 4.2, it is seen that the auto-transformer remains connected across

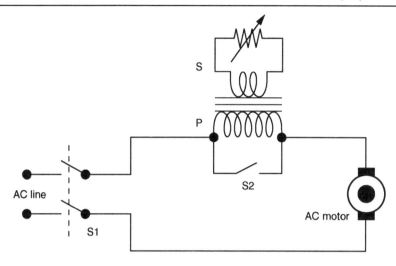

Fig. 4.3 *Reactance-type starter for AC motors. The inductive reactance of the transformer primary is maximum with the rheostat in its highest resistance setting. When the rheostat is in its zero resistance position, the residual reactance of the primary winding is near zero. Switch S2, when closed, enables the application of full line-voltage to the motor. Instead of a rheostat, tapped resistances could be used. This scheme is particularly adaptable for motors in the 1–10 horsepower range, but can be scaled up or down.*

the capacitor in the *run* mode. However, the total winding or 'secondary' displays such a high impedance that its effect is negligible.

Gentle start-up of AC motors

A simple scheme for starting single-phase AC motors is shown in Fig. 4.3. This is essentially a reactance type starter and is particularly convenient for the experimenter. It enables motor starting with greatly diminished mechanical and electrical disturbance. When used with integral horsepower motors, the common nuisances of blown fuses, activated circuit-breakers, noisy mechanical couplings and excessive light dimming can be avoided.

Current-limiting is provided by the inductance of the primary winding of the transformer. However, the actual reactance of this winding is controllable by the rheostat connected across the secondary winding. The motor starting procedure involves setting the rheostat at its highest resistance position, then closing the line switch, S1. As the motor gradually develops speed the rheostat is slowly moved to its minimum or zero-resistance position. In so doing, progressively less inductive reactance is reflected in the path of the motor current and increasing voltage is applied to the motor.

Finally, switch S2 is closed in order to impress the full line-voltage on the motor. Note that there have been no interruptions of motor current.

Some experimentation is required in such a set-up. For operator safety reasons, a step-down transformer should be selected. One capable of providing a maximum of 60–70 volts in its secondary winding should be suitable. The scheme is workable for almost any step-down ratio; the only thing affected is the resistance of the rheostat. The power rating of both the transformer and the rheostat can be less than would be required for continuous duty operation. Some sophistications can be incorporated once the appropriate components have been selected. The rheostat can be spring-loaded to return to its high resistance position. Also, the rheostat can be mechanically linked to switch S2 in order to trip this switch closed as soon as the rheostat has been manually moved to its zero resistance position.

An easy starting technique for three-phase induction motors

An interesting technique known as wye-delta starting can be used with some three-phase induction motors under certain conditions. In the interest of providing flexible options, some of these motors have the individual leads from their stator windings brought out to *six* terminals and so it is possible to connect the stator phase-windings in *either* the delta or the wye configuration. Various electronic control methods no doubt suggest themselves, but exploitation of the situation for *starting* is especially easy, and is relatively inexpensive.

Suppose that our three-phase line voltage is correct for the operation of such a motor with its stator phase-windings delta-connected. (This is the common situation for such motors.) It would be natural to consider the result of re-connecting the windings in the wye configuration. Immediately apparent is the fact that the motor would then be under-volted – each phase-winding would experience less voltage across its terminals than when participating as part of a delta pattern. Indeed, a little contemplation shows that the individual wye-connected windings receive $1/\sqrt{3}$ or about 58% of the per-phase line voltage

Recalling that torque varies inversely as the square of the line voltage and that the line current can also be expected to be reduced by 58%, a novel method of setting such a motor in operation becomes evident. All that needs be done is to provide a three-pole, double throw switch to enable switching from the wye to the delta connection of the stator windings which is what is done in the circuit of Fig. 4.4.

There is, however, a bit more than initially meets the eye. Although the application is simple enough, an ordinary knife-switch, current-capacity notwithstanding, might not prove satisfactory as it would be likely to introduce too long an interruption in the switching process. This could be

Practical projects 105

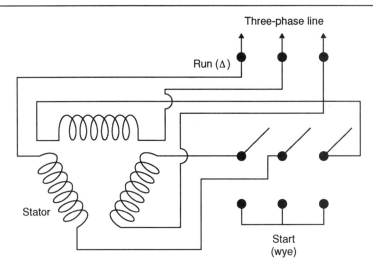

Fig. 4.4 *The Wye-delta starting method for three-phase induction motors. When the triple-pole, double-throw switch is in its* wye, *or* start *position, the phase-windings on the stator of the motor receive only 58% of their normal voltage. This provides a 'soft' start for the motor. With this switch in its* Delta (Δ), *or* run *position, the motor operates with rated-voltage applied to its stator windings. In order for the motor to have its stator windings switched from the wye to the delta configuration, six independent terminals must be provided on the stator.*

accompanied by high transient currents because of the loss of power. A spring-loaded or 'snap-action' switch is needed.

Altering the characteristics of the series motor

For some power tools and industrial process, the highly-variable speed characteristics of the DC series or AC universal motor require a bit of 'taming'. Under medium load operation, some applications would be better served if the speed did not fall off so drastically. This performance modification can be accomplished by *weakening* the field. Inasmuch as the series field also carries the armature current, the straightforward way of confining the control to the field strength is to *divert* the current around the series field winding.

This can be done by shunting the field winding with a low-resistance rheostat. See Fig. 4.5. Some judgement is needed in doing this in a safe manner, with respect to both the operator and the motor. In many cases, it would probably be best to implement the technique so that there would be no field current diversion during starting. The idea, of course, is to preserve the excellent starting torque inherent in series motors. Also, care must be given to the situation at very light loads where excessive field-weakening

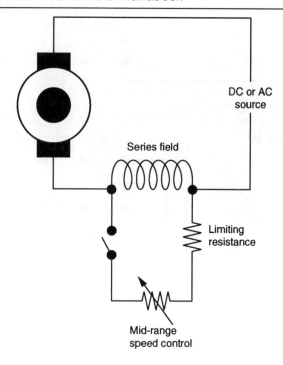

Fig. 4.5 *Method of altering the mid-range speed characteristic of series motors. The basic idea is to divert current in the series field winding, thereby weakening field-strength. This results in faster speed throughout the medium-load region, somewhat improving the rather drastic speed-regulation of series motors. The technique is also applicable to AC universal motors and to DC compound motors. Some care must be observed in implementation inasmuch as inordinately-low resistances are involved.*

could result in high-speed racing. These matters can usually be resolved with some experimentation in specific applications.

Universal motors which operate from the AC power line can be treated in basically the same way, but strict attention must be paid to safety and grounding problems. The nicest situation for both DC and AC series motors probably results when it proves feasible to incorporate a permanent shunting-resistor of such value that medium-load speed remains higher and with no discernible side effects, such as unsatisfactory starting ability, or excessive speed at light loads. With sewing machines, at least, it appears that such an overall compromise is readily attainable.

This technique is applicable to the series field of DC cumulatively-compounded motors where it provides a smooth way of controlling the speed regulation. It is particularly effective when used in conjunction with the shunt field current rheostat.

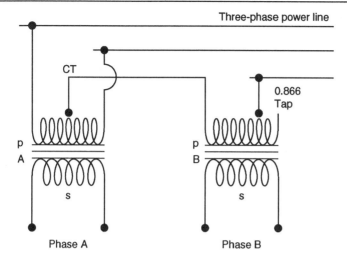

Fig. 4.6 *Phase transformation by means of the Scott connection. This arrangement of transformers converts from either three-phase to two-phase power or the converse. Transformers A and B are identical except for their primary taps. Transformer A is centre-tapped. Transformer B is tapped at 0.866 of its full-winding voltage.*

Phase transformation for motor testing

Inasmuch as most engineering laboratories are situated in industrial buildings, it is not uncommon to have three-phase power available. Altering the voltage is a straightforward job for a single three-phase transformer, or three single-phase transformers. These should be used in any event for the sake of safety and isolation. *Two-phase* power is much less common. This poses a problem when it is desired to operate and evaluate the popular two-phase induction motors used in many electrical and electronic systems. Fortunately it is possible to make use of a relatively-simple scheme long known to engineers working in the utilities industry.

The arrangement of static transformers known as the *Scott* connection is shown in Fig. 4.6. Surprisingly, this mundane circuit converts efficiently from three-phase to two-phase power and vice versa. Although the latter conversion will be less often required, it is as well to be aware of it for various experimental purposes. Those first encountering this technique, often suspect that parametric or non-linear phenomena must play a role in the phase transformation but this is not the case. Indeed the concept can be extended to allow for many types of phase transformation, such as those involving six, twelve or twenty-four phase systems. Unfortunately, the one phase conversion mode that *cannot* be achieved is the transformation of single-phase to polyphase power. This is lamentable, for if it were feasible, the low starting torque of certain split-phase induction motors could be greatly enhanced.

The 0.866 voltage tap on transformer B, known as the 'teaser' transformer is mathematically correct for nearly-perfect phase balance. It is, however, useful to realize that utilization of the full winding instead of the tap can still yield satisfactory results for many practical purposes. The practical way of reversing the phase sequence of the two-phase 'output' is to simply transpose the leads of either secondary winding of the transformers. This, of course, will reverse the direction of rotation of two-phase induction and synchronous motors.

Digitally-generated polyphase waveforms

Two-phase induction motors, including those using a capacitor for running and/or starting, achieve better phase-splitting than induction motors making use of resistance or a high-resistance winding. The first-mentioned motor works best from a true two-phase AC supply, however, all these motors conveniently operate from a single-phase line. There are, however, some *disadvantages* to capacitors. Capacitors tend to be physically large and are costly items and the electrolytic types have undesirable leakage and ageing characteristics. Temperature rise affects the capacity and the leakage as well as shortening the life-span, and as these capacitors must carry relatively high current it is not always easy to endow them with stability and long life. Oil-filled, paper and plastic capacitors tend to be inordinately bulky and serve best with smaller motors. Yet another shortcoming of capacitors is that they generally have to be *changed* if a wide speed-range is desired by varying the frequency of the AC supply.

These deficiencies can be circumvented by initially generating the polyphase format by *digital logic*. Although fancy stepped waves can be produced which simulate sine waves, simple square waveforms suffice in most practical situations. The important aspect of this technique is that the angular displacements of the phases are precise. For example, the phases of a two-phase format are exactly 90° apart *regardless of frequency* and the phases of a three-phase format are symmetrically displaced 120° from one another. After formation in the low-level logic circuits, the polyphase waves must be boosted in power amplifiers to a level suitable for motor operation.

The two-phase version of logic-generated waves is shown in Fig. 4.7. Note that the clock frequency must be *four* times the frequency to be applied to the motor. The two 'D' binaries are configured as a twisted-ring counter. The similar arrangement for generating the three-phase format is shown in Fig. 4.8. Here JK binaries are used and the clock rate must be *six* times the desired output frequency. Some care in using these circuits is necessary, as they can be triggered into locked-up states by transients. This calls for a voltage-regulated power supply and attention to bypassing.

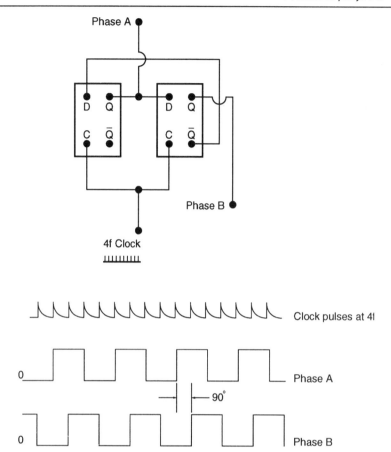

Fig. 4.7 *Basic two-phase logic circuit. The 90° time-displaced waveforms are generated by the twisted-ring logic circuit. The power-level of the waves must be boosted sufficiently to operate appropriate induction motors. The square voltage waves will produce smooth current-waves because of motor inductance. The salient feature of this two-phase format is its independence of frequency.*

Synchro-system experiments

One of the interesting techniques accompanying World War II radar technology was a non-electronic analog method of communicating angular or positional information from a sender to a remotely located receiver. This involved so-called *synchro* machines which resembled various machines now used for other purposes. It must be conceded that this scheme was inferior in both precision and elegance to modern systems based upon stepping motors

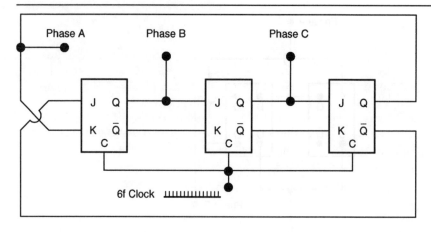

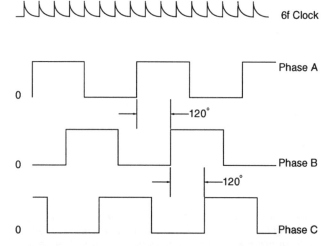

Fig. 4.8 *Basic three-phase logic circuit. The 120° symmetrically-displaced waveforms are generated. This format is independent of frequency and after an appropriate power boost, the format is suitable for powering three-phase motors. Because of motor inductance, the current waves will tend to be trapezoidal and rounded. With some additional complexity, the circuit can be rendered immune to malperformance from transients.*

and dedicated IC controllers or even servo systems using shaft-position encoders/decoders. On the other hand, the accuracy can be adequate for many purposes, reliability is excellent because of the basic simplicity of the method, and there is something psychologically intriguing about watching one machine obediently reproducing the shaft position of another.

The best way for the experimenter to investigate this phenomenon would be to acquire a pair of synchro machines, one transmitter and one receiver from the surplus market. (The two machines are very similar, but the receiver incorporates a damping mechanism to prevent excessive oscillation. For demonstration purposes, these machines can be used interchangeably. The problem is that they are not easy to find anymore.)

The stators of synchro machines resemble the stators of three-phase induction or synchronous motors, or of brushless DC motors. The rotor is a wound salient-pole type, not unlike that of some synchronous motors. A common misunderstanding is to suppose that when the rotor winding is energized with AC, three-phase voltages will appear in the stator windings. If this were true, we would have a very simple means of converting single-phase energy to the three-phase format right at the point of application. Keep in mind, however, that the induced voltages in the stator coils are generally *not* equal and are *not* displaced from one another by 120-*electrical* degrees as required in the three-phase format. Rather, the three induced voltages are always either in or out of phase with one another and all three never have the same amplitudes. In other words, these three induced 'secondary' voltages are phase-wise just like the three secondary voltages of a simple single-phase transformer with three secondaries.

Note that the stator windings of the transmitter and the receiver machines are connected in parallel. The practical result is that circulating currents will flow in the stator windings in such a way as to align the rotors of the two machines in similar angular positions. When this is achieved, the circulating currents then disappear. In a sense there is negative feedback via the self-nulling tendency through the mechanical motion of the rotor. The author has successfully used these machines for remotely rotating small beam and dish antennas.

The experimenter can obtain two similar automobile alternators and very shortly investigate the behaviour of an elementary Selsyn system, see Fig. 4.9. The basic idea is to remove the rectifying diodes (and in some cases, the built-in voltage regulator). The stator coils will probably be already connected in a 'Y' configuration, but equally good results can be obtained with the delta connection. The important thing is to have like connections in the two machines. The field windings which formerly carried DC are now energized with 50/60 Hz AC, preferably via a variable auto-transformer and during experimentation by an isolation transformer. Be careful not to apply too much voltage to the field winding as once Selsyn action is observed, excessive current in the field windings only leads to detrimental heating and is not good for the slip rings either.

If both machines are free to turn, they will share the alignment torques 'felt' by both rotors and each machine will participate in the 'following' process. In most applications, the transmitting machine will be less free to turn than the receiving machine so that the latter will re-establish its angular

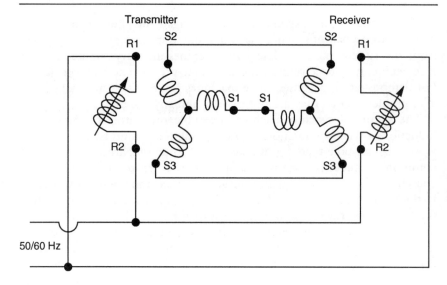

Fig. 4.9 *Simple synchro system for remotely controlling angular position. The rotors of the two machines experience electromagnetic torques which oppose any difference in their angular positions.*

If actual synchro machines are not available, fairly good results can be obtained from two similar automobile-alternators in which the rectifying diodes (and voltage regulator) are removed or disconnected. The AC voltage applied to the wound-rotors must be determined empirically. (Four instead of five interconnecting wires can be used by allowing any one of them to serve a double purpose.)

position to coincide with that of the transmitter. This, indeed, is a circuitous definition of which machine is transmitter or receiver and from purely electrical considerations, one could send positional information in either direction.

Operating AC motors from 50 Hz or 60 Hz lines

Ideally, one should supply power to AC motors only at the frequency designated on the nameplate. In actual practice, it is often satisfactory to apply 50 Hz power to a 60 Hz motor or vice versa. And some AC motors are rated for use at either 50 Hz or 60 Hz. It is difficult to make firm statements regarding power line frequency because of the closeness between 50 Hz and 60 Hz. On the other hand, motors designed for operation at either of these frequencies would be almost certain to malperform at $16\frac{2}{3}$ Hz or at 400 Hz.

There are a number of variables that enter the picture including operational duty-cycle, ventilation, starting torque, rating conservatism, speed tolerance, allowable temperature rise, power line regulation, etc. One should keep in mind that the rated-load speeds of induction and synchronous motors will differ for 50 Hz or 60 Hz lines. For example, four-pole synchronous motors, whether of 50 Hz or 60 Hz nameplate rating will rotate at 1500 rpm on 50 Hz lines and at 1800 rpm when powered from 60 Hz lines. The disparity between the speeds of induction motors operated from 50 Hz and 60 Hz lines will span approximately the same percentage spread.

When AC motors are operated at lower than their nameplate designated frequency, it is possible to obtain improvements in starting torque and in full-load torque. It may not always be wise to exploit such performance enhancement because of the inevitable temperature rise that will be sure to accompany it. Another way of looking at this is to decide whether reduced life or increased maintenance would be acceptable. Operating a 50 Hz motor at 60 Hz tends to be a more benign situation and tends to yield a worthwhile improvement in breakdown torque. It is as if the motor is slightly over-designed.

Finally, the use of motors at other than their nameplated frequencies can make them more vulnerable to the effects of low and high line voltage. This, too, translates into higher temperature rise. Providing the foregoing matters have been given due consideration, the resolution of possible problems with 50/60 Hz motors and power lines should prove fairly straightforward in most practical situations.

Changing the function of a dynamo

Engineering texts drive home the interchangeability between motors and generators. It is often mentioned that the very same machine can provide either function depending simply on whether its input is electrical or mechanical energy. This is an academically sound discourse, for it is important to understand that a dynamo always operates simultaneously as a motor and a generator. In a given instance, one function predominates over the other and we intuitively realize whether we are dealing with a motor or a generator.

Some important aspects of this interchangeability often escape specific mention. These have practical relevancy and although they could be inferred by way of study of the principles laid down in the text, they tend to be easily overlooked. In the interest of our practical work, it is best to deal with them directly.

It is easy enough to accept that a DC machine can be used either as a motor or a generator. However, suppose that we have at hand a cumulatively-compound generator and wish to deploy it as a motor. In so doing, we would probably discover that we have a strangely misbehaving motor with possible tendencies towards instability. For, whereas we had a *cumulatively*-compound generator, it is all too evident that we now have a

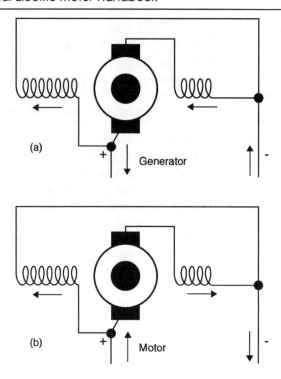

Fig. 4.10 *Interchanging machine functions affects the compounding. When the cumulatively-compound generator of (a) is operated as a motor (b), a differentially-compounded behaviour results. This is because the relative current directions in the shunt and series field windings have reversed (note arrows).*

Assuming the shunt field to be the dominant field (the usual case), the remedial measure is to transpose the leads to the series field. (In the event the main field flux was provided by the series winding, the remedy would entail transposition of the leads to the shunt winding.)

differentially-compound motor. It works both ways – a machine that is differentially-compound in one function will behave as a cumulatively-compound machine in the opposite function.

The underlying reason for this transition is that the current direction in the series field winding *relative* to that in the shunt field winding *reverses*. This is brought out in the diagrams of Fig. 4.10. Transposition of the leads of the series field winding will remedy this situation.

Worthy of mention here is the difference between the AC synchronous motor and the AC induction motor. If driven by mechanical rotation, both machines become alternators and the frequency of the driven synchronous motor is directly proportional to speed. In contrast the driven induction motor, now an induction generator, obligingly delivers current into the AC power line but at the frequency of the line *regardless* of speed. (It must be driven *above* synchronous speed, however.)

Dynamic braking of permanent magnet motors

Compelling features of the permanent magnet DC motor are its simplicity, efficiency and high starting-torque. The development of high-performance magnetic material has contributed significantly to the popularity of this motor, as earlier versions were necessarily larger and heavier and were susceptible to irreversible demagnetization, largely from the effects of armature reaction. One of the nice things about the modern permanent magnet motor is its superb behaviour where dynamic braking is needed. But here, one must observe a *caveat*.

It has been previously pointed out that the magnetism of motors and generators relying on permanent magnet fields is not 'used up'. It is as if such magnetic fields were catalytic-agents in the energy-conversion process. Nonetheless, damaging demagnetization can occur during dynamic braking.

In dynamic braking, the maximum stopping action will occur from a short-circuit of the armature. Under this condition, however, virtually all of the stored kinetic energy and inductive energy in the motor will be converted into heat dissipation in the armature windings. This may be damaging enough, but even worse, the peak current demanded from the armature can be accompanied by a sufficiently-strong magnetic field to permanently demagnetize the field magnet.

The practical remedy is to dump the motor's (now a *generator*) energy into an external load-resistance, see Fig. 4.11. This resistance should be high enough to satisfy the reasonable stopping time imposed by most practical applications. Usually nothing is gained by forcing excessively rapid deceleration, which can also be adverse to long bearing-life. Aside from a rigorous mathematical analysis with difficult-to-obtain motor parameters, a satisfactory load-resistance is usually easy to find with a little experimentation. Simply start with obviously too-high resistance and work down until a satisfactory braking-time is achieved. The required wattage of this load resistance can be inferred from its temperature rise.

Dynamic braking of induction motors

Power tools can often be made more useful by extending control of their electric motors. For those tools employing any of the several types of single-phase induction motors, a simple circuit can enable these motors to be brought to an immediate halt rather than being allowed to coast to standstill. A momentary application of direct current to the stator windings accomplishes the trick. Such braking appears to have no deleterious effect

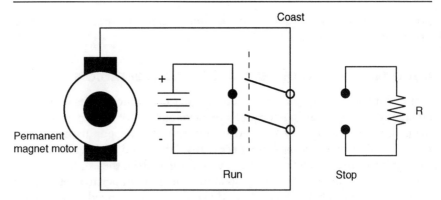

Fig. 4.11 *Set-up for dynamic braking of a permanent magnet DC motor. The coasting motor acts as a generator and dumps its rotational and inductive energy into a load resistance, R. In so-doing, it experiences a counter-torque which causes deceleration to standstill. Excessively-rapid braking because of too-low load-resistance can irreversibly demagnetize the field magnet.*

on the motor; these motors are electrically and mechanically rugged because of the rather brutal service they are expected to encounter.

The basic implementation of the technique is shown in Fig. 4.12. The switch is a SPDT type with centre OFF and one side spring-loaded. Such a switch allows the braking to be briefly applied, sparing the motor from damage due to the high current drawn by the stator. The polarity of the DC makes no difference. The diode takes some punishment both from the high surge-current and the inductive voltage-kick from the stator. A silicon rectifying-diode rated for 30 amperes and 400 volts should provide an ample safety-factor, although a small heat-sink is mandatory since there is no assurance of the exact duration of the 'momentary' braking operation. This technique will *not* work with series or universal motors; indeed, they would be likely to coast even longer. (Synchronous and shaded pole motors are worth a try.)

Another version of this braking technique is shown in Fig. 4.13. In this circuit, the energy stored in a capacitor is dumped into the stator. A practical feature here is that a more ordinary switch is used. The size of the capacitor is best determined experimentally, as the larger the capacitor the more rapid the deceleration will be. A good starting place would be several-tens of microfarads (μF) for small fractional-horsepower motors. For 0.25 HP motors and larger the capacitor may be physically larger than desired and too costly, as well. Its voltage rating should be about twice the peak voltage of the AC line. Electrically, the demand on the diode is benign, but a several-ampere unit will have the appropriate physical ruggedness. The charging resistor is not critical; a 50 K, 2 W type will serve the purpose. Note that the 'Brake' position of the switch is also the 'Off' position.

Practical projects 117

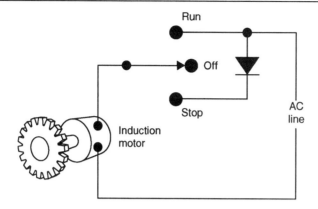

Fig. 4.12 *Braking technique for induction motors. A momentary injection of direct current into the stator causes the induction motor to come to an immediate halt. The switch is an SPDT type with the centre position OFF and the STOP position spring-loaded. A heavy-duty (30 A, 400 V) silicon rectifying-diode provides the DC braking pulse.*

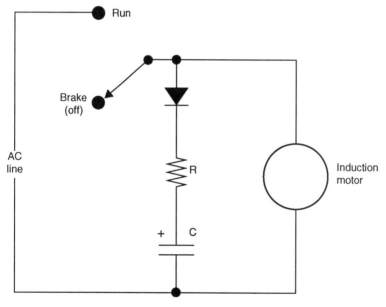

Fig. 4.13 *Alternative braking scheme for induction motors. Direct current from the charged capacitor, C is dumped into the stator winding of the induction motor. The resultant action is analogous to what would happen if a DC generator were short-circuited – immediate stoppage of the machine. Charging resistance, R can be in the vicinity of 50 KΩ, high enough so that its effect is negligible during RUN. C should be determined empirically and 30 uF would be a good start for 1/10 hp, or smaller, motors.*

Speed control of fan motors

Classical engineering texts describe the squirrel-cage induction motor (along with the DC shunt motor) as a 'constant-speed' machine. What is meant is that the speed–load regulation is very good, i.e., the speed does not fall much as more mechanical loading is applied. Of course, the *true* constant speed motor is the synchronous motor which must rotate at the synchronous rate determined by the frequency of the AC line and by the number of poles designed into the motor. These same treatises often lament the fact that the squirrel-cage induction motor does not lend itself to speed control. If one tries to obtain speed control by taking control of the applied stator voltage, say by means of a variable auto-transformer, it will be observed that the motor 'runs out of torque' before an appreciable speed reduction can be achieved.

It is nonetheless possible in the real world to obtain a speed control range of the order of three to one from a squirrel-cage motor. In so doing, however, a major concession must be made to the traditional viewpoint – the motor shaft must 'see' a soft load. In practice, such loads are provided by *fans* and *blowers*. This is not excessively restrictive, for there is great demand for systems using fans and blowers. Also, it should be mentioned that the single permanent-capacitor induction motor is the best candidate for speed control. This is because such an induction motor does not use a centrifugal switch, which might cause complications at lower speeds. However, if one is aware of this, it may prove quite feasible to employ the ensuing technique with the popular resistance-start split-phase induction motor.

The triac control circuit shown in Fig. 4.14 provides the aforementioned speed control because the peak value of the stator voltage is not changed very much, even though the rms value of stator current undergoes considerable change. Torque, being roughly proportional to the square of the peak stator-voltage retains sufficient magnitude to keep the motor operating at low speeds. It is because this torque is developed over relatively-short durations that the motor load is best restricted to fans and blowers.

Speed control with and without speed regulation

Phase-control circuits similar to those much-used in light dimmers are often used to control the speed of DC series motors or AC universal motors. These circuits usually employ a single SCR to provide half-wave unidirectional pulses to the motor. Although this imposes torque and speed limitations, the operation proves satisfactory for many applications. The SCRs are electrically rugged and are inexpensive, and the half-dozen or so accompanying components are easily available. Also, the circuit can be

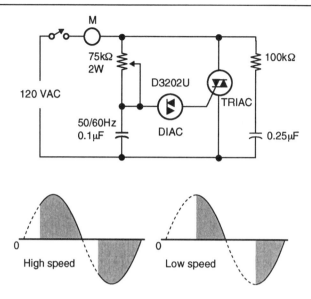

Fig. 4.14 *Circuit for speed-controlling the squirrel-cage induction motor. For best results:*
(a) Use a permanent-capacitor single-phase squirrel-cage motor. The idea is to dispense with the centrifugal switch built into other types of induction motors.
(b) The load on the motor should be either a fan or a blower.

conveniently scaled up or down to accommodate differently-rated motors. This applies to frequency as well as to torque or horsepower – 400 or 800 Hz motors can be handled pending the appropriate changes in the *RC* phase-shift network. (Other motor-types can sometimes be satisfactorily controlled by this method.) Some confusion is often seen in the technical literature between circuits such as those exemplified by Figs. 4.15 and 4.16. Despite close similarity in the configuration of these two phase-control circuits, there is a significant difference in their operation. The circuit of Fig. 4.15 enables adjustment of the speed of the motor, however, for any particular adjustment the speed will fall drastically as more load is applied to the motor. This, of course, is the 'natural' behaviour of series and universal motors. In other words, there is speed control, but *no* speed regulation.

With the slightly different circuit of Fig. 4.16, the speed is similarly adjusted by controlling the time delay of the gate-triggering voltage to the SCR. However, when the motor speed attempts to fall from increased load, the resultant reduction in motor counter-e.m.f. shortens the time delay for the gate voltage to reach triggering level. Accordingly, the motor receives a longer pulse of operating voltage (and current) which restores its speed.

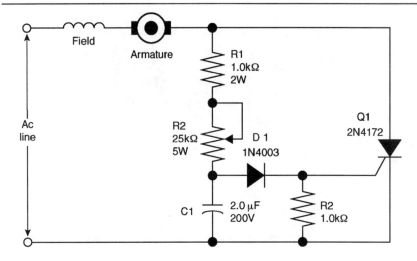

Fig. 4.15 *SCR speed-control of series and universal motors – no speed regulation. Half-wave alternations of the AC line voltage can be conducted by the SCR, but only for a duration governed by the time-delay of the gate trigger-voltage. The time-delay is adjustable by potentiometer R2, which is part of an RC phase-shift network in the gate circuit. The effective voltage applied to the motor can thereby be controlled.*

What we have is an 'invisible' feedback path which deploys the counter-e.m.f. as an 'error' signal'.

Thus, the apparently-trivial circuit differences lead to significantly different performance and cannot be ascribed to the draughtsman layout techniques.

Practical aspects of the brushless DC motor

The brushless DC motor is basically an electronically commutated dynamo, one in which the function of the erstwhile brush and commutator-assembly is replaced by solid-state drivers and electronic logic. This is much more than the long-heralded substitution of electronics for mechanical devices. Inasmuch as such parameters as timing, waveshaping and level control are easily manipulated with electronic circuits, the *tailoring* of motor characteristics becomes a reality. Many of the long enduring conflicts in selecting motor types lose their relevancy. A single machine can be driven to simulate the behaviour not only of the classical DC motors, but of induction and synchronous motors as well. Moreover, former problems with start, reversal and, of course, sparking and commutation no longer assert themselves. In general, the salient features of the brushless DC motors are:

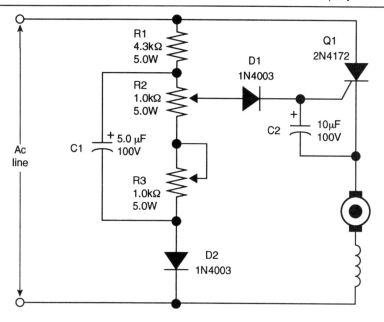

Fig. 4.16 *SCR speed-control circuit with speed regulation. Although the topography is similar to that of the circuit shown in Fig. 4.15, the small changes bring about the additional operating feature of speed regulation. In this circuit, the counter-e.m.f. of the motor affects the triggering delay-time of the SCR in such a way as to maintain near-constant speed despite variations in the load.*

(a) Much higher speeds are feasible than with brush–commutator DC motors – a typical comparison is on the order of 75 000/6500 rpm.
(b) High torque and high speed can *both* be produced and therefore a high horsepower output can be developed from a relatively small and light machine.
(c) The EMI and RFI associated with the arcing and sparking of mechanical commutation is absent.
(d) Brushless motors are safer in explosive environments – special housing or construction may not be necessary in many cases.
(e) Maintenance is greatly reduced because of the absence of commutator and brushes.
(f) Thermal problems are minimized because the external stator is the wound member.
(g) Mechanical and electrical simplicity characterizes the DC brushless motor. It comprises a *polyphase-wound stator*, a *permanent-magnet rotor*, *rotor-position sensors* and a *dedicated IC* providing the power and switching logic for driving the stator windings.

Improving performance of stepping motors

Conceptuously, it woudn't be far-fetched to suppose that the stepping motor might have been the first converter of electrical energy to rotational mechanical torque. This stems from the basic simplicity of this device. The rotor is either a permanent magnet or a ferro-magnetic member and the stator is comprised of windings arranged around its circumference. By supplying these windings with sequential pulses of appropriate nature, it is easy to visualize operation as the rotor continuously seeks new positions of magnetic alignment. Historically, the principle was not ready for practical implementation until solid-state techniques made it easy to generate the requisite logic and drive power. Moreover, computer software has played a major part in the continuing progress in the control of stepping motors.

Two winding patterns are commonly used. These are the unipolar and bipolar windings which are shown in Fig. 4.17. In the unipolar arrangement, individual coils are either on or off, i.e., the coils can assume only one magnetic polarity. In the bipolar arrangement, individual coils are switched in drive-voltage polarity and therefore in magnetic polarity. High torque/low speed requirements generally favour bipolar windings. Generally, the unipolar winding allows higher stepping rates but this may be at the expense of crispness of the stepping increments. In practice, one has considerable control over the extension of stepping rate and/or torque.

One practical improvement technique is to increase the voltage of the stepping pulses, but at the same time to maintain safe operating current by the insertion of resistances in series with the coils. In some instances, a single such resistance suffices for all coils. Performance improvement ensues because the L/R time constant of the system is thereby made *smaller*. This means that current can both rise faster and decay faster, and that a longer 'flat-top' of the drive pulse prevails. In turn, both stepping rate and torque can benefit. Of course, there is a great deal of power wasted in the resistance(s). A more elegant way to accomplish the same objective is to feed the coils a high peak-amplitude chopped wave with a duty-cycle such that the rms value of coil current remains safe.

Custom-designing of stepping motors

The performance enhancement methods just described pertain to a stepping motor which you already have. However, if an order hasn't already been placed with the distributor or manufacturer, there may be another option for obtaining a higher stepping rate, more torque or both. It consists of querying the maker as to whether a given stator size can be wound with coils of more than just one wire gauge. If so, the solution could be to ask for windings with heavy-gauge wire. Such a winding will have relatively low resistance and also low inductance. The latter parameter falls off very rapidly

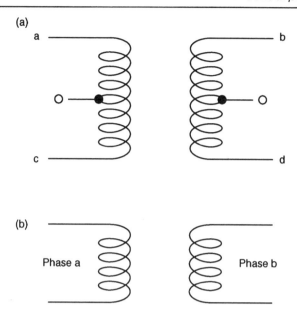

Fig. 4.17 *The two common winding-patterns of the stators of stepping motors.*
(a) Unipolar: The coils are often bifillar-wound with the centre-tap brought out. The basic idea is that only one coil with respect to centre-tap is energized at a time. No polarity reversal takes place.
(b) Bipolar: Here, the sequencing logic involved in the stepping increments makes use of polarity reversal in the coils. Push-pull or H-drivers are needed.

as larger wire is substituted for smaller wire. This is because inductance is proportional to the *square* of the number of turns. It may appear that the lower resistance of heavy gauge windings contradicts the previously described technique where resistance was deliberately *added* to the stator windings in order to improve performance. However, the decreased inductance of the heavy gauge coils is much greater than the decreased resistance. The net effect, therefore, is a *reduction* in the L/R time constant. This, again leads to a higher stepping rate. It also leads to greater torque, but now for *two* reasons.

The improvement in torque stems from the reduction of the L/R time constant and also from the greater current in the heavy gauge windings. Although a given torque rating is largely based upon the number of ampere-turns developed in the windings, it is often a practical fact that a few turns of heavy wire can have a higher *allowable* current than might be expected from simple ampere-turn calculations. In some systems the *need* for higher operating current may be a problem.

The bottom line is that the stepping motor provides accurate and repeatable positioning as a simple open-loop device. Servo-systems based upon *conventional* motors are much more complex and tend to suffer from the instabilities that plague feedback systems and moreover they are more costly because they require tachometers, encoders and decoders, up–down digital counters and other specialized circuit-blocks.

Electronic technique for eliminating the centrifugal switch

If one were looking for shortcomings in the wide variety of single-phase induction motors, the centrifugally-operated switch would be a prime candidate. This mechanically-actuated device enables the motor to develop starting torque by one means or another and then, when the motor attains high speed, the switching transition occurs so that operation ensues as a simple induction motor. This is acceptable but lower maintenance would be achieved if the centrifugal mechanism could be *eliminated*.

The experimenter can benefit from a practical technique that has served well with garage-door systems. The simplified diagram of Fig. 4.18 shows how the direction of rotation of a two-phase type AC motor is electronically reversed. By means of appropriate logic, one or the other Triacs is triggered into conduction. This scheme thus dispenses with a mechanically-actuated switch. The basic idea of this set-up can be adapted to replace the centrifugal switch associated with other AC single-phase induction motors.

The capacitor-start induction motor shown in Fig. 4.19 uses a Triac to connect and disconnect the starting-winding circuit. Triggering of the Triac is controlled by a current-transformer which senses the current drawn by the running winding. This current is quite high at standstill and low speeds. Initially, therefore, the Triac *conducts*, completing the AC circuit to the capacitor and the starting winding. As the motor gains speed, the current in the running winding and in the primary of the current transformer naturally declines until at some point in the acceleration, the Triac will no longer be triggered into conduction and the starting winding circuit will be *opened*. This, of course, simulates the function of the conventional centrifugal switch. Some version of this scheme should be applicable to the other single-phase induction motors presently relying on the centrifugal switch, and obviously, some empirical effort will be required to optimize performance.

Some motor drive techniques

The control of motors has become largely a matter of selecting an appropriate IC module and then providing the requisite boost of power level. Indeed, if the motor isn't too large, application is often simplified via a single IC. Nonetheless, it is useful for the experimenter to have at hand various

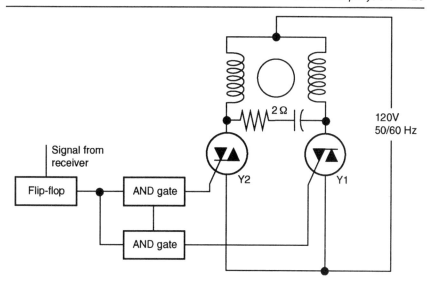

Fig. 4.18 *Electronic reversing technique used in garage-door systems. The two-phase type induction motor rotates in one direction or the other depending upon which Triac is being triggered into conduction.*

quickly-implementable techniques for conveniently putting motors through their paces.

Bipolar transistors in the Darlington configuration are often used in motor drive circuits. It can prove useful to be aware that there are actually *four* circuit arrangements qualifying as Darlington amplifying stages. (Keep in mind, too, that monolithically-integrated Darlingtons characterized for motor drive are commonly available.) Fig. 4.20 shows the four discrete-element Darlington circuits.

A cascade-connected pair of discrete transistors can be readily put together to simulate the salient characteristics of the IGBT. In Fig. 4.21, the input transistor is a small MOSFET and the output device is a power bipolar transistor. Together, they provide high-impedance input and low voltage-drop to the load. Pioneered and advocated by the International Rectifier Corporation, this circuit is known as the BIMOS switch and, as its name suggests, it is particularly useful in switching-type control systems. An example would be PWM control of armature current in DC motors.

Control of small motors is, largely for economic reasons, often accomplished with a single phase-controlled SCR. This is essentially similar to the circuitry found in the low-cost light-dimmers. The half-wave drive is not ideal for either DC or universal motors and smoother operation and extended speed range would be forthcoming from a full-wave control waveform. Fig. 4.22 depicts an easy way to get full wave control from a single

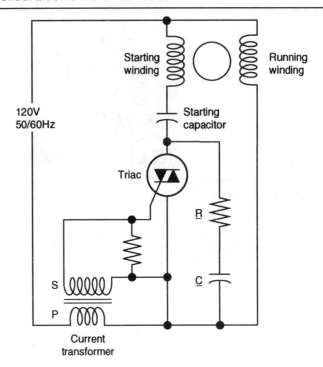

Fig. 4.19 *Electronic replacement circuit for the centrifugal switch. Several types of single-phase induction motors have* starting *and* running *windings. The starting winding is opened by a centrifugal switch when the motor attains speed. This is done by the Triac in the above arrangement. The primary of the current transformer has one to several turns of heavy conductor and does not degrade motor operation. The* RC *'snubber' network can be* 100Ω *and* 0.01 μF.

SCR. In some applications this scheme because of its electrical ruggedness, merits choice over a Triac. Triacs for the control of 400 Hz motors tend to have marginal ratings, making them vulnerable to failure with modest combinations of temperature and electrical stress. Note that the SCR, diode-bridge combination fits directly into control circuitry that would otherwise use a Triac.

A nice control technique for small universal motors is readily provided by a phase-controlled Triac. The full-wave format enables smooth control down to low speeds. The basic control circuitry resembles that used in light dimmers and is both low cost and simple. A problem arises in extending this technique to larger universal motors because of the unavailability of higher power-handling Triacs. A neat trick is to simulate Triac operation with two SCRs. These are connected in anti-parallel, as shown in Fig. 4.23. The heart of the circuit is the three-winding transformer. For all practical purposes, the input or primary winding of this transformer functions as the gate of a Triac.

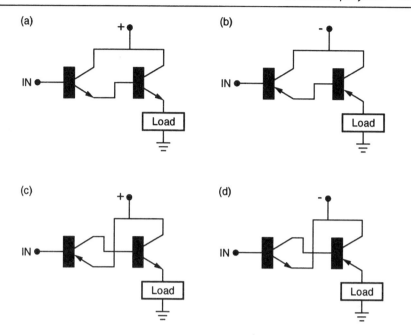

Fig. 4.20 *Four Darlington stages using discrete bipolar transistors. In all circuits, the input transistor is a low-level type and the output transistor is a power type.*
(a) Conventional NPN circuit.
(b) Conventional PNP circuit.
(c) Simulated PNP circuit.
(d) Simulated NPN circuit.

What is essentially required are three conductively-isolated windings with a three or four step-down ratio from primary to secondaries, which are identical. Many common pulse-transformers meet this non-critical requirement.

Note the phasing of the secondary windings is such that positive trigger signals are simultaneously delivered to the gates of the two SCRs. Nonetheless, the SCRs fire *alternately*, not together because only one SCR at a time receives positive anode voltage from the AC line. There is no limit with regard to power and safety factor, frequency is no problem and the SCRs provide high current and a high voltage capability at modest cost.

This type of full-wave control is particularly good for experimentation with fan motors. The various single-phase induction motors can undergo a surprising amount of speed variation despite their reputation for being insensitive to applied voltage and a small speed change can have a significant effect on fan performance. Also, one recognizes that high starting torque is

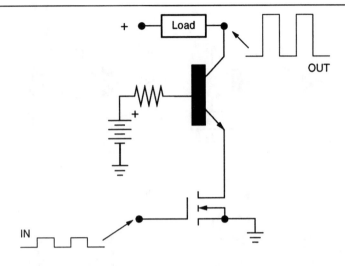

Fig. 4.21 *The BIMOS switching circuit. This discrete driver-circuit is comprised of a small MOSFET input stage and a power bipolar output stage. Overall, the circuit has high input impedance and the common-base output transistor features much better voltage and frequency capability than is forthcoming from conventional common emitter stages. Additionally, the low voltage-drop simulates the behaviour of an IGBT device. The battery signifies the required bias provision for the bipolar transistor.*

not a usual requirement for fans. Conceivably, low-maintenance operation could be obtained from a permanent-capacitor induction motor or a large shaded-pole motor because neither has brushes or a centrifugal switch. Keep in mind that the chopped waveform, provided by the anti-parallel SCRs causes a low power-factor in the AC line *regardless* of motor parameters.

IC control system for permanent magnet DC motors

Many semiconductor firms market many motor–controller ICs. These often incorporate a host of 'bells and whistles' that would be difficult or costly to implement via the older design approach of using discrete devices. It usually suffices to view the IC as a 'black-box', i.e., one doesn't have to know very much about the internal electronic circuitry beyond basic capabilities and pin-connections.

The motor-control set-up shown in Fig. 4.24 exemplifies this situation. This control system includes several compelling features:

(1) Speed control of a 0.5 HP permanent magnet DC motor is provided. The basic circuit lends itself to easy scale-up or scale-down to handle other motor sizes.

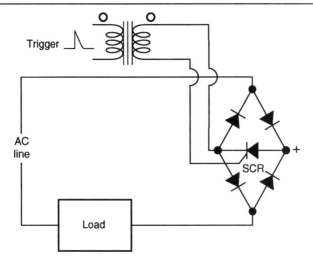

Fig. 4.22 *Technique for obtaining full-wave control from a single SCR. The 1:1 pulse transformer is often unnecessary, but helps avoid ground conflicts during experimentation. The circuit simulates the operation of a Triac, but has much greater power and frequency capability. It is useful, for example in working with 400 Hz aircraft motors. Conventional rectifying diodes suffice for the bridge.*

(2) The commonly-tolerated power loss in the feedback sampling resistance is negligible here.
(3) Transformerless operation directly from the AC line is provided.
(4) Control is attained via a PWM wave at about a 2 kHz repetition rate. This obviates the need for expensive or fragile high-frequency components.
(5) The IC automatically starts the motor in a soft-start mode.
(6) Over-current protection is incorporated in the IC.

A word is in order with regard to the unique power-MOSFET. Note the extra pin-connection to the so-called 'mirror' element. Involved here are just a few of the thousands of cells comprising the drain-source section of the device. At only a few milliamperes, the mirror circuit faithfully delivers a scaled-down version of the 10 or more amperes of peak current in the drain circuit. Thus, the mirror element is essentially another *source*, one particularly suitable for instrumentation purposes. Instead of requiring a five or 10 W feedback sensing-resistance, a one-quarter watt rating does the trick here. Motorola, the maker, calls this power-device the *SenseFET*. Aside from its extra current-sensing provision, it exhibits the usual features of power-MOSFET technology – easy drive, high-frequency response and relative immunity to thermal-runaway.

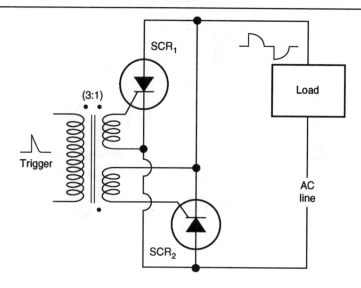

Fig. 4.23 *Another simulated Triac circuit for full-wave control of AC motors. The arrangement comprises two SCRs connected in anti-parallel. The SCRs conduct alternately at delay times governed by the single trigger pulse. Unlike triac devices, there are virtually no practical power or frequency limitations. The circuit is usually incorporated in a feedback system. The full-wave control is particularly effective with universal motors. Other AC motors can be speed-controlled through a smaller range if they are associated with fan-type loads. (Use a bridge-rectifier ahead of a DC motor.)*

IC control system for brushless DC motor

Because of its mechanical simplicity and its amenability to electronic control, the brushless DC motor has rapidly been gaining in popularity. Deployment of this motor has been greatly enhanced by the availability of dedicated ICs which conveniently allow control of the motor's characteristics and consequently some of the traditional controversies and trade-offs pertaining to the choice of the most suitable motor have lost their relevancy. Desirable features of both AC and DC motors are combined in both design and operation. Structurally, a permanent magnet rotor, a polyphase stator, and stator-position sensors are involved. The sensors may be optical or magnetic types and frequently, they are Hall-effect types. Commutation consists of supplying appropriately-timed pulses to the stator windings. Unlike, the stepping motor, the rotor provides a smooth torque rather than being incrementally jerked through its rotational cycle.

The set-up shown in Fig. 4.25 makes use of bipolar power-transistors to drive the motor. With minor changes, other devices such as power MOSFETs, IGBTs, or Darlingtons could be used and Y-connected stator-coils

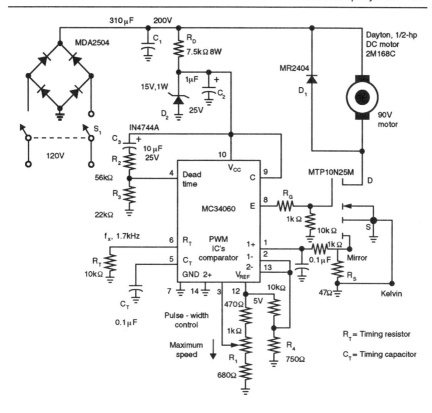

Fig. 4.24 *IC control-circuit for a permanent magnet DC motor. The SenseFET power device functions similarly to conventional power MOSFETs, but virtually eliminates power dissipation in the feedback-sensing resistance. Despite its high immunity to thermal runaway, an appropriate heat-sink should be provided. A good starting point would be approximately 16 sq. inches of 1/8" aluminium, or equivalent. Inasmuch as there is no transformer isolation from the AC line, care is needed in order to avoid ground conflicts and shock hazards.*

can also be used, keeping in mind that a higher operating voltage is then needed.

An additional IC is required for this control system. This is a pulse-width modulator which is a readily available IC because of its widespread application to regulated power-supplies. The Motorola MC3420 and MC3520 ICs are well-suited to the task. The pulse-repetition rate is not critical and the upper audio frequencies or low supersonic frequencies have commonly been used. In operation, the effective motor current and therefore the speed is controlled by the duty-cycle of the pulse-width-modulated control signal. This scheme preserves the torque capability down to very-low speeds. The low-speed 'cogging' that often plagues low-speed operation of DC motors

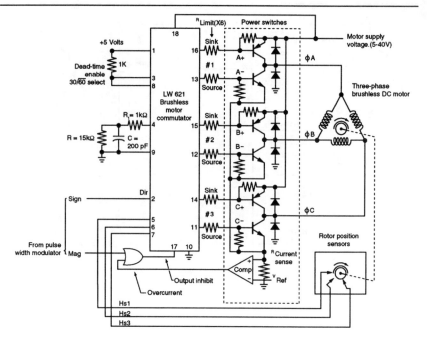

Fig. 4.25 *Commutation of a three-phase brushless DC motor. The National Semiconductor LM621 IC electronically accomplishes the primary function of the mechanical brush and commutator system. Additionally, the operating characteristics of the motor can be varied at will. An additional IC, a pulse-width modulator is required to complete the control system. Manufacturer's literature for the LM621 explains how to deal with motors having different numbers of phases and with different sensor spacing.*

is not a problem with this technique. The logic sign at pin 2 governs the direction of rotation – no connection change in the motor circuit is needed. The physical resemblance of these motors to conventional three-phase synchronous motors suggests a path for experimentation. Conversion of the AC synchronous motor to the brushless DC motor would require sensors to supply commutating data to the control IC.

IC energy-saving system for induction motors

Despite the widespread use of the induction motor in industry and in the home, a decidedly-adverse characteristic of this otherwise superb motor was often tolerated as being unavoidable. This pertains to its low power factor and its resultant low efficiency at light loads. Investigation revealed a means of automatically lowering applied voltage at light loads would still enable the motor to carry that load, but at a *higher power factor and at lower line-current*. With the advent of solid-state electronics and especially integrated-circuit modules, a practical and cost-effective way of accomplishing this became available.

This self-adjustment of the light-load power factor distinguishes induction motor behaviour from that of a transformer. The transformer also manifests a low, lagging power factor at light loads, but is not amenable to much change in power factor by a reduction of voltage to its primary winding. Both the induction motor and the transformer appear more resistive and less inductive at or near full load, thereby drawing line current at a good (high) power factor. The idea behind an induction motor energy saver is to bring about high power factor via lowered voltage (together with reduced current-consumption) for a *lightly-loaded* operation. This saves energy, reduces operating cost, and promotes motor-longevity.

A simple system for accomplishing such enhanced behaviour is shown in Fig. 4.26. The Harris Induction Motor Energy Saver IC must be selected with regard to the intended line voltage. The HV-1000 is for 120 V use and the HV-1000A is for use with 240 V lines. A 50/60 Hz supply can be accommodated. The set-up is suitable for 1/2 HP motors and the potentiometer provides adjustment of the full-load power factor. The Triac should be sized for about three-times anticipated full-load current and should be heat-sinked.

Keep in mind that the energy-saving benefit of this technique is primarily realized in applications where the motor would ordinarily spend appreciable time in lightly loaded operation. The technique is only applicable to induction motors and appears best-suited to single-phase systems.

DC permanent magnet motor for electric vehicles

Although there is continuing controversy over whether DC or AC motors are more ideally-suited for electric vehicle propulsion, the practical fact is that many trade-offs are involved in the comparison. Additionally, the philosophies of the hobbyist/experimenter and the large manufacturer can be expected to be based upon different considerations. The availability of solid-state technology, inverters and pulse-modulation techniques means that the control of DC motors is no longer dependent upon energy-wasting rheostats. This enables DC systems to compete on a nip and tuck basis with the much-heralded efficiency of AC systems. In particular, the permanent magnet DC motor merits consideration as a cost-effective and easily-implementable source of traction power.

Extraordinary progress in the ceramic, ferrite and alloy materials used in permanent magnets is reflected in the high starting torque, small physical volume, high operating-efficiency and moderate cost of modern permanent magnet motors and many suitable types and ratings are available. One cannot go wrong in selecting this motor for the electric vehicle. Earlier experimenters based their designs on the DC series motor largely because of

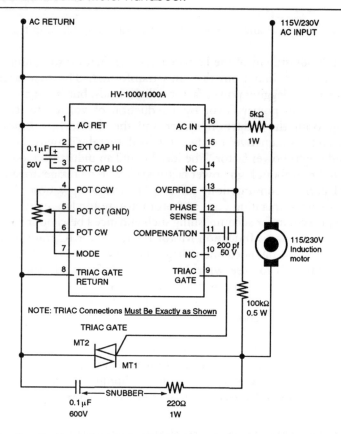

Fig. 4.26 *Application of the induction motor energy saver. At light loads, an induction motor operates at low power-factor, draws relatively high current and thereby wastes energy. If, however, the applied voltage is reduced the motor raises its power-factor in order to develop the torque it needs. This is accompanied by a reduced current-consumption because the motor now appears more resistive and less inductive. The IC module senses the lightly-loaded operation and automatically reduces the r.m.s. voltage applied to the motor. Reduced voltage and current, together with improved power factor, cut the energy wastage of motors spending appreciable time lightly-loaded.*

the availability of such machines on the surplus market. Although DC series motors have long been used for electric traction applications, the more 'civilized' torque–speed characteristics of the modern permanent magnet motor compels its consideration.

A well-performing electrified car for the experimenter or hobbyist could utilize a 15-horsepower permanent magnet motor with one or several-hundred volt rating. (It is not feasible to energize vehicle motors from low voltage because of the very high I^2R loss incurred.) Small four-cylinder 'stick-shift' automobiles generally qualify for the conversion, and for simplicity it is acceptable to retain both transmission and clutch. Don't overlook the possibility of selling the internal combustion engine to help soften the economic impact of the conversion. Control ICs and power electronics are abundantly marketed by the major semiconductor firms.

Switched reluctance motor

The type of motor shown in Fig. 4.27 uses a soft-iron rotor and develops rotational torque from sequential switching of the stator poles. There are those who expect such a motor to seriously threaten the widespread popularity of the induction motor in the near future. This may appear a strange prognostication in view of the fact that one sees no new basis for motor action in this configuration. Indeed, we are quickly reminded of stepping motors, brushless DC motors, and even the venerable synchronous motor. Moreover, a bit of historical research reveals that the underlying concept of continuous rotation is a very old one that, until recently, could not efficiently and economically emerge as a practical machine.

Now, due to the availability of both, IC logic-modules and a variety of solid-state power devices, these 'switched reluctance' motors merit renewed investigation. The power devices include the power MOSFET, the IGBT, bipolar transistors and Darlingtons, GTOs and other thyristors and power op-amps. These can provide current-capabilities of tens, hundreds, and even thousands of amperes. At the same time, voltage ratings in the hundreds and thousands of volts have become common. Best of all mass-produced devices at such previously-unthinkable ratings can be had at surprisingly low-cost.

The salient benefits of these switched reluctance motors include ease of manufacture, no permanent magnet, electronics 'tailoring' of performance parameters, operation from any type of power-line and the possibility of integral containment of the drive and logic electronics. Of course, brushlessness remains a compelling feature.

Minimal machine-shop equipment should be needed for the experimenter to construct such motors. For the most part components from conventional motors should ease the burden of translating design to hardware. A great deal of flexibility in physical pattern is allowable because motor performance will largely be the task of electronic logic. The laminated soft-iron or silicon-steel rotor in Fig. 4.27 has six-poles, whereas the stator has eight. This results in smoother running torque, and in a starting torque that is less dependent upon the standstill position of the rotor. Possible candidates for control logic are ICs such as the National Semiconductor LM621, and the Motorola MC33035.

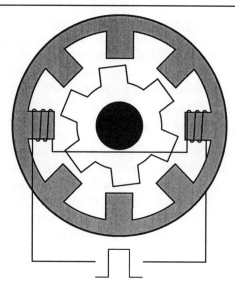

Dc pulses for one of
four-phases of stator excitation

Fig. 4.27 The switched reluctance motor. First cousin to a stepping motor, the design and operation emphasizes power and torque rather than rotor-position control. Rotation takes place from the appropriately-sequenced stator pulses because of the tendency of magnetic flux to seek an alignment of minimum reluctance. The rotor is made of laminated 'soft' magnetic material and does not become permanently magnetized. The drawing shows one of the four phases, or pole-pairs, that receive the timed DC pulses.

Reliability

Rapid turn-on and turn-off of solid-state power devices has dramatically enhanced the overall performance of switching power supplies and inverters used for driving motors. The resultant improvement in the efficiency of motor systems has, however, proved a mixed blessing in some instances, for, despite continuing progress in metallurgy, bearing precision and electrical insulation those concerned with motor reliability and maintenance have observed a somewhat mysterious increase in *both* bearing and insulation failure during the past decade. In practice, however, it hasn't required a Sherlock Holmes to suspect a connection between the rapid switching transitions and the motor reliability problems; the defects have been too repetitive.

The near-instantaneous switching transients, although short-lived, are heavily invested with high-frequency harmonic energy. This is exacerbated in integral horsepower systems where high currents are handled. Shock-excitation of stray resonances takes place and the radio-frequency energy tends to *concentrate* at points or over small physical areas. Very much higher currents or voltages can exist in restricted regions rather than elsewhere in the system. High current concentrations account for the thermal pitting of the bearings. High voltage concentrations can account for insulation destruction, especially at the very beginnings of stator windings and spot-heating also occurs.

It is difficult to prescribe cures for individual situations. Various grounding, filtering and bypassing techniques can prove helpful. If a reduction in radiated EMI or RFI around the motor is observed this may be an indication of reduction in shock-excited resonances. Other things being equal a short connecting cable to the motor can reduce the severity of reflected energy from transmission-line resonances. It is not easy to postulate exactly how the radio-frequency energy finds its way into the bearings, although radiation and/or stray coupling are the likely culprits.

It may prove safer to drive motors from supplies in which internal switching rates don't exceed the 'old-fashioned' 20 kHz rate. The trade-off of greater size and weight for the avant-garde power supply design could prove a wise compromise in the light of the rather nebulous remedies against damage from high-frequency energy. Motors operated directly from the power line should die of old age rather than from other causes.

5 Practical problems

Dealing with motor mathematics

The ensuing problems have been selected to illuminate the motor behaviour discussed in the previous chapters. Although the qualitative and the quantitative approaches each stand on their own, the *combination* of the two should provide greatly-enhanced insights into the whys and wherefores of motor performance. Most motor mathematics makes use of very-ordinary algebra, with a smattering of trigonometry and geometry thrown in. Sometimes the arithmetical operations can prove tedious; ironically, it is here that it will likely be rewarding to make haste slowly. Some of us, ability notwithstanding, lack the patient temperament of good accountants.

Of course, all the caution and self-checking in the world are no compensation for a problem illogically set up. The prime procedural admonition is therefore to understand the nature of the problem. In many motor problems, quantities not specifically asked for, must nevertheless be derived in order to ultimately work out the answer. This chain of sub-answers can be no stronger than its weakest link. And, it should go without saying that cause-and-effect relationships must always be apropos to the sought goal.

All-in-all, these problems are devoid of trickery and trivia and representative of the practical aspects of electric motors.

Feeder line as part of the motor circuit

PROBLEM: The following problem may not appear directly relevant to electric motors, but the practical fact is that the power feeder line is very much part of the motor's circuit. This is commonly manifested by an unsatisfactory starting torque, worse-than-expected speed regulation, and sometimes, abnormal temperature rise.

A 15 HP, 240 volt DC motor operates at a full-load efficiency of 87%. The motor starting-current is 1.5 times the rated full-load current. The motor is located 600 feet from the DC supply. It has been determined that the maximum voltage drop in the feeder during starting is 25 V. What cross-sectional area of copper wire will just meet this requirement? (assume zero voltage-drop in the DC supply.)

SOLUTION: Motor full-load current,

$$I_L = \frac{746 \times HP}{V \times \text{Efficiency}} = \frac{746 \times 15}{240 \times 0.87} = 53.6 \text{ A}$$

Motor starting current = $1.5 \times 53.6 = 80.4$ A

Overall feeder line resistance $= \dfrac{25}{80.4} = 0.311 \, \Omega$

Minimum cross-sectional area $= \dfrac{10.4 \times 600 \times 2}{0.311} = $ **40 129 Circular Mills**

Use # 4 copper wire

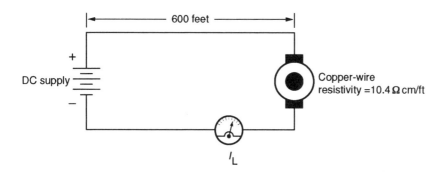

Internal power in a permanent magnet DC motor

PROBLEM: An engineering laboratory wanted to determine the full-load internal horsepower of a permanent magnet DC motor being developed for electric vehicle propulsion. Normally, one has to measure the armature resistance in order to calculate the counter-e.m.f., which in turn, enables the calculation of the internally-developed power.

Because of the combined error sources comprised of skin-effect (armatures of DC motors carry AC), brush–commutator voltage-drop, temperature-rise and instrumentation difficulties it was decided to set up a test procedure to circumvent the need for direct measurement of armature resistance.

To accomplish this, the motor was operated as a generator feeding the same electrical load it would have consumed from the power line when performing as a motor. An adjustable speed drive motor provided the proper full-load speed. The motor-turned-generator delivered 120 V at 100 A. What internal horsepower did this signify?

PROCEDURE:
(1) The key to this problem is the realization that the generated output voltage corresponds to the counter-e.m.f. with the machine operated in normal fashion as a *motor*.
(2) The internal wattage of a motor is the product of counter-e.m.f. and armature current.
(3) Make use of the relationship, one horsepower = 746 W.

SOLUTION:
Internal wattage = 120 × 100 = 12 000 W or 12 kW
Conversion to horsepower = 12 000/746 = **16.1 HP**

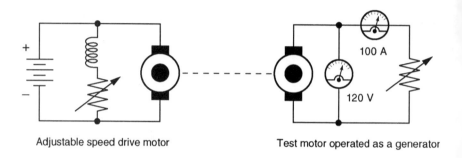

Adjustable speed drive motor Test motor operated as a generator

Stray power in a permanent magnet DC motor

PROBLEM: The no-load power input to a motor consists largely of the frictional and windage losses, usually lumped together as stray losses. Providing this evaluation is not affected by abnormal operation, such as runaway or by residual loading from reduction gears, stray power measurement helps designers choose better bearings and provide more-effective ventilation. A 100 V, 20 HP DC permanent magnet motor draws 3.25 A at no-load. What is the stray-power?

SOLUTION: The power consumed in stray dissipation is closely approximated by the no-load wattage input.
Stray Power = Applied Volts × Resultant Current = 100 × 3.25 = **325 W**

COMMENT: 325 watts is the equivalent of 325/746 or 0.44 HP. This is of the order of 2% of the rated output, which is reasonable.

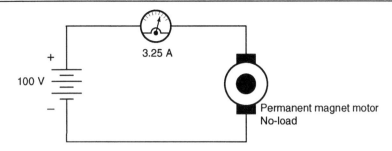

Power determination from prony-brake measurements

PROBLEM: Although it requires a bit of practical skill, the prony-brake technique of measuring mechanical output of a motor has been widely used. Consider a 220 V DC motor that draws 20 amperes at full load. A prony brake with a two-foot radius arm provides a corrected reading of 12 lb at a moment when the speed is recorded as 1050 r.p.m. From this simple set-up calculate the torque, horsepower output, electrical input power, and the efficiency of the motor.

PROCEDURE: The prony brake enables determination of the torque. This, in conjunction with the speed leads to evaluation of the horsepower output. The electrical input power is the product of line voltage and line current. From the foregoing, sufficient data is available to set up the output/input ratio representing efficiency. *Both* output and input must be expressed in either mechanical or electrical units.

SOLUTION: Torque = Prony brake radius arm × force = 2 × 12 = 24 lb-ft

$$\text{Horsepower out} = \frac{2\pi \times \text{Torque} \times \text{Speed}}{33\,000} = \frac{2\pi \times 24 \times 1050}{33\,000} = \mathbf{4.80\,HP}$$

Electrical input power = line voltage × line current = 220 × 20 = **4400 W**

$$\text{Efficiency} = \text{output/input} = \frac{4.83 \times 746}{4400} = 0.81 = \mathbf{81\%}$$

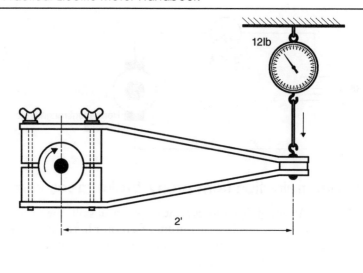

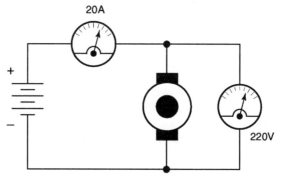

Compound motor speed calculations

PROBLEM: In the engineering laboratory of a training facility a simple set-up is retained to demonstrate the effects of compounding DC motors. A DC compound motor with free-access armature and field terminals has the following characteristics.

Armature resistance, $0.5\,\Omega$. Series-field resistance, $0.2\,\Omega$. Brush voltage-drop, 3 V (constant). Power source, 100 V DC.

At one-quarter full-load, line current is 5 A as a shunt motor; its speed is 1500 r.p.m.

Assume that insertion of the series-field can either increase or decrease the magnetic flux in the air-gap by 10%, depending upon whether cumulative or differential compounding is brought about.

Calculate full-load speeds for (a) shunt-motor operation, (b) cumulatively-compounded operation, (c) differentially-compounded operation.

SOLUTION: The basic relationship involved is $S = \text{cemf}/k\phi$ or $k\phi = \text{Cemf}/S$

In more useful form for our problem: $S = \dfrac{E_a - [I_a R_a + I_a R_s] = BD}{k\phi}$

where ϕ is magnetic flux in air-gap
 k is a constant
 E_a is armature voltage
 R_a is armature resistance
 R_s is series-field resistance
 I_a is armature current
 S is speed
 BD is brush volt-drop

$$k\phi = \dfrac{1 - [5 \times 0.5 + 5 \times 0.2 + 3]}{1500} = 0.0623$$

(a) Shunt full-load speed $= \dfrac{100 - [20 \times 0.5 + 20 + 3]}{0.0623} = 87/0.0623 = \textbf{1396 r.p.m.}$

(b) Cumulative compound, $S = \dfrac{100 - [20 \times 0.5 + 20 \times 0.2 + 3]}{0.0623 \times 1.1} = 83/0.0685 = \textbf{1212 r.p.m.}$

(c) Differential compounds $S = \dfrac{100 - [20 \times 0.5 + 20 \times 0.2 + 3]}{0.0623 \times 0.9} = 83/0.0561 = \textbf{1479.5 r.p.m.}$

COMMENT: Too much differential compounding results in instability.

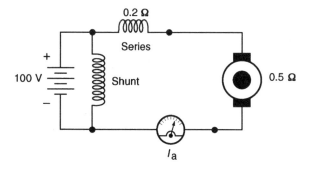

Speed regulation of DC and AC motors

PROBLEM: For a certain application, the speed regulation of two selected motors are to be compared. One is a DC shunt motor with 6% speed regulation. The other motor is a 50 Hz single-phase four-pole induction motor. The AC motor is specified to have a 6% slip and its no-load speed has been determined to be 1455 r.p.m. In making the comparison, the speed at no-load of the DC motor will also be adjusted to be 1455 r.p.m. Which motor will exhibit the best (lowest) speed regulation?

PROCEDURE: In dealing with the concepts of speed regulation in the DC motor and with slip in the AC motor, we have to be careful we do not compare apples and oranges. The logical approach is to manipulate the slip information so that the speed change in the AC motor can be expressed in the *same way* as for the DC motor.

SOLUTION:

The synchronous speed of the AC motor is $\frac{120 \times 50}{4} = 1500$ r.p.m.

Full-load speed of the DC motor is $(1 - 0.06)(1455) = 1386$ r.p.m.

Full-load speed of the AC motor is $(1 - 0.06)(1500) = 1410$ r.p.m.

For *both* motors, speed regulation $= \frac{\text{no-load speed} - \text{full-load speed}}{\text{full-load speed}} \times 100$

For the AC motor, speed regulation $= \frac{1455 - 1410}{1410} \times 100 = \mathbf{3.2\%}$

The AC induction motor shows the better (lower) speed regulation.

Note: Approximations inherent in this problem do not affect the basic conclusion.

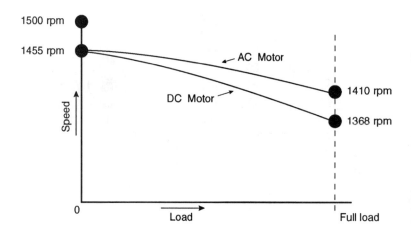

Torque calculation of DC series motor

PROBLEM: A DC series motor develops 100 lb-ft of torque when its line current is 20 A. What torque can be expected when the motor draws 30 A?

PROCEDURE: With the limited data given, it must be assumed that the motor continues to operate within the linear region of its magnetization curve throughout the 2–30 A current range.

The basic torque relationship is $T = kI_a$ where k is a fixed constant, ϕ is the air-gap flux, I_a is armature current.

SOLUTION: We do not know ϕ, but ϕ is proportional to armature current over the linear portion of the magnetization curve. It is therefore possible to make the following simplification:

$$T = k(I_a)^2$$

Inasmuch as we will be interested in just setting up a proportion k can be given the value 1.

Then the equation expressing the proportional relationship will be:

$$T = 100 \times \frac{30 \times 30}{20 \times 20} = 225 \text{ lb-ft}$$

Efficiency of DC shunt motor

PROBLEM: Determination of the full-load efficiency of a 100 volt DC shunt motor is required. Full-load line-current is 5 amperes. The armature resistance is 0.5 Ω. The field resistance is 200 ohms. The overall mechanical losses comprising bearing friction, brush friction and windage amount to 74 watts. From manufacturer's data, a reasonable estimate of individual brush-resistance is 1 ohm. (There are two brushes.)

PROCEDURE: Our objective will be to sum up the losses and then make use of the relationship:

$$\text{Per cent efficiency} = \frac{\text{input} - \text{losses}}{\text{input}} \times 100$$

SOLUTION: Copper loss in field winding $= \frac{(100)^2}{200} = 50$ W

Field current $= \frac{100}{200} \times 0.5$ A $= 0.25$

Armature current $= 5 - 0.5 = 4.5$ A

Copper loss in armature $= (4.5)^2 \times (0.5) = 10.13$ W

Stated mechanical losses $= 75$ W

Electrical loss of two brushes $= 2 \times (4.5)^2 \times 1 = 40.5$ W

Summation of all losses $= 50 + 10.13 + 40.5 + 75 = 175.63$ W

Per cent efficiency $= \frac{500 - 175.63}{500} = \frac{324.37}{500} = 65\%$

COMMENT: Design improvement would try to lower brush and mechanical losses.

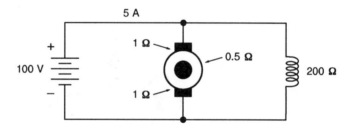

Starting resistance for permanent magnet DC motor

PROBLEM: The rated full-load armature current for a DC permanent magnet motor is 20 amps. It normally operates from a 50 V DC source. The armature resistance of this motor is 0.5 Ω before any temperature rise sets in. The brush voltage-drop can be assumed to be 2 V. The initial starting current is therefore $\frac{50-2}{0.5} = 96$ A. This is considered excessively high and comes about because at standstill there is no counter-e.m.f. to oppose the inrush of current. What value of starting resistance should momentarily be inserted in the line to limit the starting current to 150% of the rated full-load current?

SOLUTION: Making use of Ohm's law, we can set up a relationship representing the desired situation as follows:

$$R_s = \frac{V_a - (E_c + V_b)}{I_a} - R_a,$$

but E_c is *zero* at start. Therefore, we can simplify to:

$$R_s = \frac{V_a - V_b}{1.5 \times I_a} - R_a$$

$$R_s = \frac{50 - 2}{1.5 \times 20} - 0.5 = \mathbf{1.1 \ \Omega}$$

where E_c is counter-e.m.f.
       V_a is operating voltage
       V_b is brush-drop
       I_a is full-load current
       R_s is start resistance

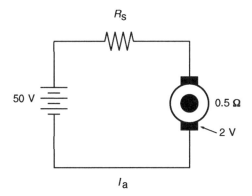

Power requirement from electric vehicle motor

PROBLEM: In the design of an electric vehicle, it has been determined that the drive-shaft will require a torque of 50 lb-ft at a speed of 1500 rpm. From this consideration alone, what must be the horsepower output of the drive-motor?

SOLUTION: From physics texts, the basic relationship is expressed by the equation:

$$HP = \frac{2\pi \times T \times S}{33\,000}$$

where T represents torque expressed in pound-feet
where S is speed in revolutions-per-minute

Cancellation yields a simpler and more familiar form of this equation:

$$HP = \frac{T \times S}{5252}$$

In our case,

$$HP = \frac{50 \times 1500}{5252} = 14.28 \text{ Horsepower}$$

In practice, one would specify a 15 HP motor. The type of motor would have to be determined by other factors, such as starting torque, costs and maintenance.

Speed vs. load for DC shunt motor

PROBLEM: A DC shunt motor is powered from a 100 V DC source.
The armature resistance is $0.4\,\Omega$.
The field winding resistance is $50\,\Omega$.
At full-load, the total line-current is 25 A.
Full-load speed is 2000 r.p.m.
(a) What is the speed at half-load?
(b) What is the speed at 125% of full-load?

SOLUTION: Full-load calculations:

$$I_a = I_1 - I_f = 25 - \frac{100}{50} = 23 \text{ A}$$

$$E_c = V_a - (I_a R_a) = 100 - (23 \times 0.4) = 90.8 \text{ V}$$

Needed data: at the rated speed of 2000 r.p.m.

$$E_c = 90.8 \text{ V and } I_a = 23 \text{ A}$$

We now have the needed data to calculate the speed of half load:

$$I_a = \frac{23}{2} = 11.5 \text{ A}$$

$$E_c = V_a - I_a R_a = 100 - (11.5 \times 0.4) = 95.4 \text{ V}$$

$$S = \text{Original speed} \left(\frac{\text{Final } E_c}{\text{Original } E_c}\right) = 200\left(\frac{95.4}{90.8}\right) = 2101 \text{ r.p.m. at half-load.}$$

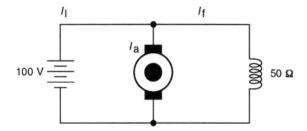

where I_1 is line current
I_a is armature current
I_f is field current
V_a is armature voltage
V_c is counter-e.m.f.

(b) In a similar manner, we can now solve for the speed at 125% of full-load:

$$I_a = \frac{5}{4} \times 23 = 28.75 \text{ A}$$

$$E_c = 100 - (28.75 \times 0.4) = 88.5 \text{ V}$$

$$S\frac{5}{4} = \left(\frac{88.5}{90.8}\right) \times 2000 = \mathbf{1949 \text{ r.p.m. at } 125\% \text{ full-load}}$$

COMMENTS: The above calculations represent a good practical procedure. For greater precision, one would need to take into account the voltage drop across the brushes, as well as other elusive losses.

Speed vs. load for DC series motor

PROBLEM: A DC series motor is powered from a 240 V DC source.
The armature resistance is 0.1 Ω.
Brush voltage-drop is assumed to be 3 V.
The series field resistance is 0.05 Ω.
When the load causes the line current to be 80 A, the measured speed is 500 r.p.m.
(a) What is the speed when the load is increased to 100 A?
(b) What is the speed when the load is decreased to 50 A?

SOLUTION: Calculate the counter-e.m.f.s (E_c) for the three loads. Use $E_c = V_a - I_a(R_a - R_s) - V_b$

$$E_{c1} = 240 - 80 \times (0.1 + 0.05) - 3 = 225 \text{ V at 80 A load.}$$

$$E_{c2} = 240 - 100 \times (0.15) - 3 = 222 \text{ V at higher load}$$

$$E_{c3} = 240 - 50 \times (0.15) - 3 = 229.5 \text{ V at lower load}$$

Proportional equivalents will now be used. These will involve E_c, and ϕ. On the assumption that field strength is proportional to armature current, we can simply set up the proportions using I_a in place of ϕ.

$$S_2 = S_1 \times \left(\frac{E_{c2}}{E_{c1}}\right) \times \left(\frac{I_{a1}}{I_{a2}}\right) = 500 \times \left(\frac{222}{225}\right) \times \left(\frac{80}{100}\right) = 395 \text{ r.p.m. at 100 A load.}$$

$$500 \left(\frac{229.5}{225}\right)\left(\frac{80}{50}\right) = 816 \text{ r.p.m. at 50 A load.}$$

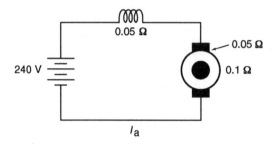

where I_a is armature, line, and load current.
E_{cx} represents counter-e.m.f.
S_x represents speed.

This motor is operated over the linear portion of its magnetization curve. ϕ is the air-gap field strength.

Synchronous speed calculations

PROBLEM: The control system for an industrial processing project requires the same performance on both 50 Hz and 60 Hz power lines. Translated into practice, this means that two synchronous motors are needed in order to provide the same synchronous speed from the two power lines. One motor will be used with the 50 Hz line, the other with the 60 Hz line. Obviously these motors will have different numbers of poles in order to yield the *same* speed. Specify the minimum number of poles for each.

PROCEDURE: (1) Note that *any* speed is allowable just so it is identical when one motor is powered from the 50 Hz line and the other receives its power from the 60 Hz line.

(2) Make use of the relationship;

$$S = \frac{120 \times f}{P}$$

where f is frequency in Hz; P is number of poles and S is speed in r.p.m.

(3) *P* must be divisible by 2, as fields are produced by pairs of poles.
(4) To arrive at the minimum number of poles, make a table commencing with two poles. Increase the number of poles until a common speed can be attained by selected 50 Hz and 60 Hz motors.

SOLUTION:

No. of poles	Speed of 60 Hz motor	Speed of 50 Hz motor
2	3600	3000
4	1800	1500
6	1200	1000
8	900	750
10	720	600
12	600	500

Inspection reveals that a 10-pole 50 Hz synchronous motor and a 12-pole 60 Hz motor will each provide the synchronous speed of 600 r.p.m. when connected to its appropriate power line.

Light-load behaviour of induction motor

PROBLEM: In this problem, the numbers are contrived in a manner to drive home an important aspect of motor operation. However, the performance described is within the bounds of reason.

A 10 horsepower induction motor has a full-load efficiency of 90.5% and its power-factor is also 90.5%. This motor operates from a single-phase 230 V power line. Measurements taken for half-load operation show that both efficiency and power-factor have dropped to 64%. What are (a) the full-load and (b) half-load line currents?

PROCEDURE: The mathematics of this problem are straightforward; we must convert the mechanical output power to its electrical equivalent, then solve for line current by applying Ohm's law as it pertains to power, i.e., $I = P/E$. The trick is to properly modify the relationship by taking into account the efficiencies and power factors.

SOLUTION: Electrical output power = $10 \times 746 = 7460$ W.

(a) Full-load line-current = $\dfrac{7460}{230 \times 0.905 \times 0.905} = \mathbf{39.6\ A}$

(b) Half-load line current = $\dfrac{3730}{230 \times 0.64 \times 0.64} = \mathbf{9.6\ A}$

COMMENT: The fact that full-load and half-load line currents 'happened' to be the same in this problem demonstrates the *wastage of energy* typical of lightly-loaded induction motors. It is also seen that the required line current is affected in the same way by both efficiency and power factor.

Using wattmeter data for evaluating induction motor performance

PROBLEM: A 1/2 HP single-phase induction motor operates at full-load from a 50 Hz power line. Three instruments, a wattmeter, a voltmeter, and an ammeter are used for taking electrical data. The recorded readings are 450 W, 120 V, and 5 A. What are the power-factor and the efficiency of the motor under these conditions?

PROCEDURE: The power factor must be determined *before* attempting to calculate the efficiency from the relationship:

$$\text{Efficiency} = \frac{HP \times 746}{E \times I \times PF}$$

Here, the dimensional equivalent is output/input and the factor 746 converts horsepower to watts.

SOLUTION:

$$PF = \frac{450}{120 \times 3} \text{ (from power factor = true power/apparent power)} = \mathbf{0.75}$$

$$\text{Efficiency} = \frac{0.5 \times 746}{120 \times 5 \times 0.75} = \mathbf{0.83 \text{ or } 83\%}$$

COMMENT: A deceptive advertising ploy used for certain equipments (notably, power supplies) is to boast about the *high* efficiency, but remain silent about the *low* power factor. With multi-horsepower induction motors, power factor is of prime importance to the utilities company, and to all parties concerned with industrial economics.

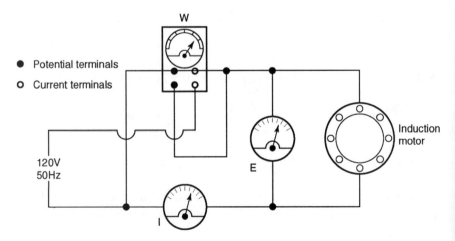

Speed of wound-rotor induction motor vs. rotor resistance

PROBLEM: The per-phase resistance of the rotor-coils in an eight-pole wound-rotor induction motor is $0.2\,\Omega$. The rotor terminals are short-circuited and the motor is powered by a 60 Hz three-phase line. The full-load speed is then measured to be 850 r.p.m.

If the short-circuit is replaced by $0.8\,\Omega$ resistance per phase, what will be the new full-load speed? (Assume 'Y'-connected rotor-coils and 'Y'-connected resistances.)

PROCEDURE: First, it is necessary to determine the synchronous speed. Next, making use of the fact that slip is proportional to rotor resistance, a proportion can be set up relating the *new* slip to the *original* slip with shorted rotor-terminals. Finally, the above ratio can be translated into the new speed with the added rotor resistances.

SOLUTION: The synchronous speed is

$$\frac{120 \times f}{P} = \frac{120 \times 60}{8} = 900\ \text{r.p.m.}$$

Note: Because we are dealing with a three-phase motor, it should be clear that '8-poles' implies 8-poles per phase, or a total of 24 poles on the stator. (A *total* of 8 poles would be an impossible construction.)

$$\text{Original slip} = \frac{900 - 850}{900} = 0.0167$$

$$\text{New slip} = \frac{1.0}{0.2} \times 0.0167 = 0.0835$$

New speed $900 \times (1 - 0.0835) = 900 \times 0.9165 = \mathbf{824.9\ r.p.m.}$

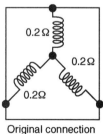

Original connection

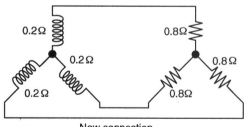

New connection

Using 50 Hz motors on 60 Hz lines and vice-versa

PROBLEM: A 50 Hz induction motor is to be used on a 60 Hz line of the same voltage. Inasmuch as horsepower is directly proportional to speed, it was anticipated that the motor would deliver about 60/50, or 20% more horsepower. However, this proved to be far from the case. What might have been overlooked?

SOLUTION: The notion of a 20% increase in output horsepower derives from the basic equation:

$$HP = \frac{2\pi \times T \times N}{33\,000}$$

where T is torque in lb-ft.
where N is speed in r.p.m.
where 33 000 represents the number of foot-pounds per minute in one-horsepower.

The probable weak point in the expectation of a 20% gain in output power is the *reduction* in torque from 60 Hz operation. This stems from the fact that the applied voltage requires increasing by 20% to compensate for the 20% increase in inductive reactance for 60 Hz operation.

This, of course, is an approximate approach which neglects other factors, such as increased friction, windage, hysteresis and eddy-current losses. It is, however, representative of adjustable-speed systems which endeavour to obtain a speed range from an induction motor without *also* varying the applied voltage from a variable-frequency power-supply.

COMMENTS: This problem has been presented to stimulate instructional insight. The operation of electric motors beyond their intended electrical and mechanical ratings requires serious consideration of the effects on maintenance, longevity, and safety.

Interpreting data for polyphase motors

PROBLEM: Initial inspection of two synchronous motor stators does not reveal differences in their windings. Students are informed that one is intended for a single-phase line and the other for a three-phase line. Also, both stators will provide six poles when energized. What will be the synchronous speed of the two motors?

PROCEDURE: The speed of the rotating magnetic field of such stators is readily calculated by the handbook formula,

$$S = \frac{120 \times f}{P}$$

where S is the speed in r.p.m.
       f is the frequency in Hz.
       P is the number of poles.

SOLUTION: For the single-phase motor:

$$S = \frac{120 \times 60}{6} = 1200 \text{ r.p.m.}$$

From the stated data, this *might* also be the speed of the three-phase motor. There is a lack of clarity here. By 'number of poles', it is entirely reasonable to infer that what is implied is *six-poles per phase*. In that case, the three-phase motor would actually operate at a synchronous speed of 1200 r.p.m.

COMMENT: An equally-valid argument, however, could construe that 'number of poles' to mean the *total* on the stator. In that case, there would be two-poles per phase and the synchronous speed would be 3600 r.p.m.

(Note that the stator is the *armature* of these motors.) Handbooks would avoid this confusion if they used the uncontroversial language, *'number of poles per phase'*.

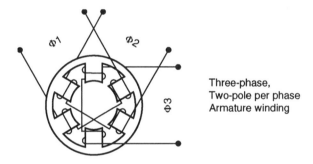

Three-phase,
Two-pole per phase
Armature winding

Capacity calculation for unity power factor

PROBLEM: A 5 HP single-phase induction motor operates from a 220 V, 50 Hz power line. The efficiency is 80% at a power-factor of 0.85. It is desired to increase the power-factor to unity by connecting a capacitor in parallel with the motor. What size should this capacitor be?

PRELIMINARY: Great precision is not required in these situations for two reasons. First, the practical capacitors have wide capacity tolerances. Second, in practice it is desirable to present the power line with a load having a *near-unity* power factor; efforts expended in attaining perfection yield negligible reward.

NEEDED DATA:

$$\text{Motor current} = \frac{HP \times 746}{V \times \text{Eff.} \times PF} = \frac{5 \times 746}{220 \times 0.85 \times 0.80} = 25 \text{ A}$$

Inasmuch as an induction motor is an inductive reactance and causes a *lagging* current, we seek cancellation by the *leading* current of a capacitor.

SOLUTION: Capacitor current = $(25) \times (\sin \theta)$, where θ is the angle whose cosine is 0.85. Thus, θ is 31.7 degrees and the corresponding sine is 0.53 (from trigonometric tables)

Capacitor current = $25 \times 0.53 = 13.25$ A

$$\text{Capacitive reactance} = \frac{220}{13.25} = 16.6 \, \Omega = X_c$$

$$\text{Capacity in } \mu F = \frac{10^6}{2\pi \times 50 \times 16.6} = 192 \, \mu F$$

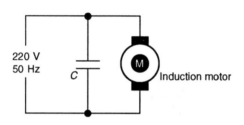

Synchronous motor calculations for power factor correction

PROBLEM: The induction motors in a factory comprise a load of 600 kW at a 0.65 power factor. An additional mechanical load of 100 HP is contemplated. It is felt that an appropriately-sized and operated synchronous motor could not only supply the additional mechanical load, but could also simultaneously improve the overall power-factor to 0.9. Calculate the required kW output and kVA input to be provided by the synchronous motor.

PROCEDURE: The procedure is straightforward when carried out on a step-by-step basis.

(1) Conversion of the stated horsepower mechanical load to kW yields the needed power-output of the synchronous motor = $746 \times 100 = \mathbf{74.6\,kW}$.
(2) Prior to installation of the synchronous motor, and before addition of the 100 HP load, the factory's kVA rating was 600/0.65 = 923 kVA.

(3) When the 100 HP load is added, the factory's power consumption will be increased from 600 kW to 674.6 kW. At the new power-level of 674.6 kW the corresponding kVA rating at 0.9 power factor will be 674.6/0.9 = 749.6 kVA.

(4) Using the numbers derived above, the solution consists of comparing the reactive powers (kVARs) of the original situation with that of the modified situation incorporating the added load and the synchronous motor.

SOLUTION:

Reactive power of original situation = $\sqrt{(923)^2 - (600)^2}$ = 701.4 kVAR

Reactive power of new situation = $\sqrt{(749.6)^2 - (674.6)^2}$ = 327.1 kVAR

kVAR to be supplied by synchronous motor = 701.4 − 327.1 = **374.3 kVAR**

kVA to be supplied by synchronous motor = $\sqrt{(374.3)^2 + (74.6)^2}$ = **381.5 kVA**

Note: The efficiency of the synchronous motor has negligible effect.

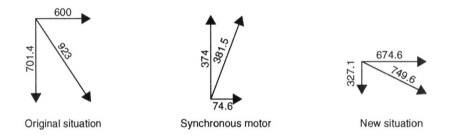

Original situation · Synchronous motor · New situation

Transformer simulation of induction motor

PROBLEM: For design and test purposes, induction motor data can be obtained by treating such motors as transformers. A 5 HP, three-phase induction motor draws a full-load line current of 17 A. With rotor *blocked*, the armature current is adjusted with a variable auto-transformer to its full-load 17 A value. The corresponding power-input is recorded at 825 W. With the rotor *unblocked*, the no-load power input is recorded as 325 W, with accompanying line-current of 7 A.

(a) What is the per-phase resistance revealed by the full-load simulation?
(b) What are the rotational losses of this motor?

PROCEDURE: (a) The per-phase resistance (between any two-lines) is 2/3 of the Ohm's law derived relationship $R = P/I^2$. Accordingly, we must deal with the value given by $R_e = 0.67 \times P_c / I_1^2$

Where, R_e is the per-phase resistance.
  P_c is the power dissipation due to combined copper losses in both rotor and stator.
  I_1 is the per-phase (line) current.

(b) The rotational losses are represented by the unloaded power-input minus the no-load copper losses.

SOLUTION:
(a) $R_e = 0.67 \times 825/(17)^2 = \mathbf{1.9\, \Omega}$
(b) Rotational losses $= 325 - (7)^2 \times 1.9 = 325 - (49 \times 1.9) = \mathbf{231.9\, W}$

COMMENTS: Interestingly, it was *not* necessary to know whether wye or delta connections were involved in the stator-windings.

Also, the calculated rotational-losses are assumed to be *fixed* losses.

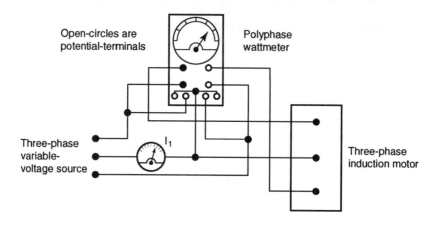

Two-wattmeter data for induction motor calculations

PROBLEM: Measurements are recorded for a lightly-loaded three-phase induction motor using the two-wattmeter method. If W1 is found to be negative 200 W, while W2 is positive 1950 W, what is the input power and the power-factor?

PROCEDURE: A lightly-loaded induction motor can be expected to exhibit a low lagging power-factor. The likelihood of such performance is, indeed, suggested by the negative value of W1.

The solutions are straightforward provided the indicated sums and differences are manipulated *algebraically* to take into account the negative sign of W1.

Input power is the *algebraic* sum of W1 and W2 = $-200 + 1950 = \mathbf{1750\, W}$.

To find the power-factor, it is necessary to first solve for tangent θ and then look up the cosine corresponding to angle θ. (By definition, power-factor is cosine θ.)

$$\tan\theta = \sqrt{3}\times\left(\frac{W1-W2}{W1+W2}\right) = \sqrt{3}\times\left(\frac{-200-1950}{-200+1950}\right) = 1.73\times\left(\frac{-2150}{+1750}\right) = -2.126$$

−2.126 is found to be the tangent of approximately −65 degrees.

The cosine of −65 degrees yields the **power factor of 0.423 lagging**.

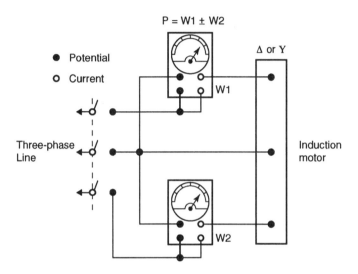

Calculations for induction–generator action from a motor

PROBLEM: The hydro-energy from a stream is to be used to produce single-phase 60 Hz power from a three-phase induction motor operating as an induction generator. No sources of electrical power are nearby so the implemented generator will have to provide its own excitation. It is known that this can be accomplished by connecting appropriate capacitors across the delta-connected windings of the stator.

The 3 HP motor requires 230 V. As a four-pole machine. Its synchronous speed at 60 Hz is 1800 r.p.m. It is also known that its no-load line current is 7 A. Approximately, what value capacitors should be connected across each phase?

PROCEDURE: This appears to be a 'fuzzy' situation inasmuch as load power will only be extracted from one of the phases. This imbalance will be dealt with later. To get started, it should be recognized that this is essentially a 'resonance' problem, similar to power factor problems. Basically, the job of the capacitors is to resonate with the magnetizing reactance of the machine. The *practical* way of proceeding is to also recognize that the no-load apparent power should be matched by the reactive power of the capacitors. This is an indirect way of manipulating parameters, but it works.

SOLUTION:

(1) Total no-load apparent powers $= \sqrt{3} \times E_{line} \times I_{line} = 1.73 \times 230 \times 7$ = **2785 VA**

(2) Per-phase apparent, or reactive power $= 2785/3 =$ **928 VAR** (good assumption)

(3) $I_{phase} = \dfrac{I_{line}}{\sqrt{3}} = \dfrac{7}{1.73} =$ **4.05 A**

(4) $C_{phase} = \dfrac{I_{phase}}{2\pi \times f \times (E_{phase})} = \dfrac{4.05}{2\pi \times 60 \times 230} =$ **46.7 µF**

COMMENT: Nigel Smith, author of 'Motors as Generators for Micro-Hydro Power', has determined that the capacitors should be deployed as shown, rather than symmetrically associated with the three-phases. This expedient promotes an artificial balance over a reasonably-wide load range. There will, however, be a favoured direction of rotation.

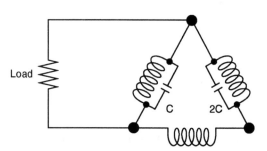

C = 46.7 µF, 2C = 93.4 µF
The exact value of these capacitors is not critical. They must be non-electrolytic types with ample voltage rating. (One phase has no capacitor.)

PROBLEM: A speed-selection system used in railroad technology is known as concatenation. Traction torque derives from two induction motors having a common shaft. By appropriate switching of the motor leads, different speeds are available. Consider an arrangement in which wound-rotor motor No. 1 has four poles and induction motor No. 2 has twelve poles. Power is taken from a three-phase 60 Hz line. What shaft speeds can be selected? (For simplicity, speed-selecting switches are not shown in diagram.)

PROCEDURE: This is a 'brain-teaser' that does not readily yield to intuitional analysis. However, a simple equation enables determination of the possible speeds. The trick is to solve for the slip, S_1, in motor No. 1:

$$S_1 = \frac{P_2}{P_2 \pm P_1}$$

where S_1 is the slip in motor No. 1
P_1 is the number of poles in motor No. 1
P_2 is the number of poles in motor No. 2.

The slip in motor No. 1 is then easily converted into the mutual shaft-speed as follows:

Shaft speed = $(1 - S_1) \times$ (synchronous speed of motor No. 1)

SOLUTION:

Synchronous speed of motor No. 1 *alone* = $\dfrac{120f}{P} = \dfrac{120 \times 60}{4} = $ **1800 r.p.m.**

Synchronous speed of motor No. 2 *alone* = $\dfrac{120f}{P} = \dfrac{120 \times 60}{12} = $ **600 r.p.m.**

S_1 (with motor No. 2 aiding motor No. 1) = $\dfrac{12}{12 + 4} = 0.75$

Speed of motor No. 1 = $(1 - 0.75) \times (1800) = 0.25 \times (1800) = $ **450 r.p.m.**

S_1 (with motor no. 2 *resisting* motor No. 1) = $\dfrac{12}{12 - 4} = 1.5$

Speed of motor No. 1 = $(1 - 1.5) \times (1800) = (-0.5) \times (1800) = $ **−900 r.p.m.**

COMMENTS: (a) Since the machines are not synchronous types, the speeds are *approximate*.
(b) The negative speed-sign denotes *reverse* rotation.
(c) Motor No. 1 operates simultaneously as a *frequency converter*.
(d) Opposite rotation of the computed speeds is also available.
(e) Auxiliary starting is needed on reverse-rotation mode.

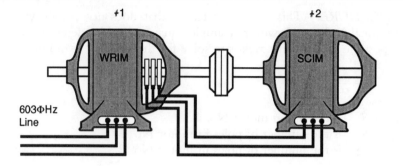

Index

AC induction motors 11, 30, 114
  as alternator 30
  counter e.m.f. 20–22
  double squirrel-cage 80–82
  IC energy-saving system 132–3
  light-load calculations 151
  low squirrel-cage resistance 63
  performance calculation 152
  polyphase 66
  shaded pole 72–3
  single-phase 63–5, 66–8
  speed regulation 118
  split-phase starting 65–6
  squirrel-cage 75–6, 118
  three-phase 11, 86–7, 94–6, 104
  transformer simulation calculation 157–8
  variant frequency calculation 154
  wattmeter data calculations 158–9
  wound-rotor 76–80, 153
AC motors 53, 62
  feedback mechanism 31
  gentle starting procedure 103–104
  non-sinusoidal waveforms 91–4
  rotation reversal 88–91
  running at non-designated frequencies 112–13, 154
  starting 99–100, 103–104
  universal operation 53, 106
  unusual behaviour 96–8
  see also under individual types

AC synchronous motors 68–72, 114
  constant speed 118
  induction type 98
  overloading 97–8
  power factor correction calculations 156–7
  speed relationship 16–17
  three-phase 11
AC/DC motors, dual functions 60–61, 87–8
Amplidynes 35
Armature (definition) 9, 11
Armature reaction 36–7, 55–6
Armature/line current 8–9
Auto-transformer, for starting 102
Automobiles, starter motor 53

Batteries, for electric vehicles 27, 29
BIMOS switch 125
Bipolar transistors 125
Braking techniques 29–30, 115–16
Brush-commutator systems 32

Capacitors
  for phase splitting 67–8
  size calculation for unity power factor 155
  for starting motors 101–102, 108
Centrifugal switch, elimination 124
Cogging 131

164 Index

Compensating windings 56
Compound motors 57
  DC 53–4
  speed calculations 142–3
Concatenation system 160
Consequent-pole motors 83–4
Control
  DC PM motors 128–9
  full-wave format 126–7
  small motors 125
  small universal motors 126
Conventional current 17
Counter e.m.f. 20–22
  practical use 57
Current flow direction 17
Current-limiting technique 31

Damper windings 70
Darlington configuration 125
Davenport, Thomas 32
DC brushless motors 27, 120–21
  IC control 130–32
  three-phase 11
DC compound motors 53–4
  speed calculations 142–3
DC motors 22, 32
  basic set-up 4–5
  feedback mechanism 31
  induced AC voltage 34–6
  reversing rotation 56–7
  speed regulation 144
  speed relationships 44–7
  starting 99–100
  see also under individual types
DC permanent magnet (PM) motors 42–3
  control 59–60
  dynamic braking 115
  IC control 128–9
  internal power calculation 139–40
  runaway immunity 43
  starting resistance calculation 146–7
  stray power calculation 140
DC series motors 26, 32, 51–3

  altering characteristics 105–106
  speed calculations 149–50
  torque calculation 145
DC servo motors 47–9
DC shunt motors 39–42
  efficiency calculation 145–6
  instability/runaway 41
  reversibility 40
  speed calculations 148–9
DC tachometer 60
DC watthour meter, motor characteristic 49–51
DC/AC inverters 30
Distortion, of air-gap field 36–7
Double squirrel-cage motor 80–82
Dynamic braking
  DC PM motors 115
  induction motors 115–16
Dynamo see Generator

Early discoveries 2–3
Electric motors
  classification features 12
  as energy converter 7–8
  energy-flow diagram 24
  nomenclature 9–11
  performance descriptions 13
  performance graphs 8–9
  symbols and schematics 13–15
  see also under individual types
Electric vehicles 26–31
  parallel hybrid type 29
  power requirement calculation 147–8
  series hybrid type 29
  use of DC PM motor 133–5
Electron current 17
Energy conversion 7–8
  idealized 23–4
Energy (definition) 8, 13
Energy-saving system, integrated circuit (IC) 132–3

Fan and blower motors, speed control

Index

118–20
Faraday disc 4, 32
Faraday, Michael 1
Feeder line calculation, as part of motor circuit 138–9
Foot-pounds units 13
Frequency, operation at non-designated 112–13
Frequency changer 78–80

Generator
  cumulatively-compound 113
  use as motor 113–14

Hall-effect sensors 130
'Hand rules' (for motor parameters) 17–18
Harris Induction Motor Energy Saver IC 133
Henry, Joseph 1
Homopolar motor 33–4
Horsepower rating 11–12, 23
Hysteresis motors 73–5

Induction generator 22, 97, 114
  calculation 159–60
Integrated circuit (IC)
  energy-saving system 132–3
  Harris Induction Motor Energy Saver 133
  motor-controller 128–9, 130–32
  pulse-width modulator 131
  reliability 136
  selection 124–5
International Rectifier Corporation 125
Inverters, DC to AC variable frequency 30

Lenz, Heinrich 1
Lenz's law 1, 6, 11
Lightly-loaded operation
  energy saving 132–3
  induction motor calculations 151

Magnet hydrodynamic generator 34
Magnetic motor 22
Magnetism, residual see Residual magnetism
Measurements, realistic 26
Mirror circuit 129
Monopole 82
Motor circuit, feeder line calculation 138–9
Motor and generator combined behaviour 54–6
Motor testing, phase transformation 107–108
Motor-action, basic 5–7
Motor-drive techniques 124–8
Motorola
  ICs 131, 135
  SenseFET device 129

National Semiconductor, ICs 135
non-sinusoidal waveforms, AC motors 91–4

Oersted, Hans Christian 1
Orthogonal relationship (force/current/flux) 18–19

Perpetual motion quest 22
Phase splitting 66–8
Phase transformation, for motor testing 107–108
Plugging 57, 88
Polyphase motors, speed calculation 154–5
Polyphase waveforms 108
  generated by digital logic 108
Pound-feet units 13
Power
  consumption in three-phase motors 94–6
  definition 8, 13
  determination from prony-brake measurements 141
Power factor 31

capacitor size calculation 155–6
correction calculations 156–7
determination 94–6
improving 69, 72
industrial economics 152
light-load 133, 151
Power feeder, as part of motor circuit 138–9
Power-MOSFET 129
Prony-brake measurements, power determination calculation 141
Pull-in/pull-out torque 72
Pulse-width modulator 131

Radar technology 109
Radio-frequency interference 137
Reactance, variable 68
Regenerative braking 29–30
Regulex 35
Reluctance motors 75–6
  switched 135
Repulsion motors 11, 35, 89–91
Residual magnetism 38–9
Resistance measurement 26
Resonance 160
Rosenberg generator 35
Rotary motion, continuous 3–4
Rotor, wound 77
Rotor (definition) 10
Rotor resistance 153
Rotorol 35

Scott transformer arrangement 107
Selsyn systems 77–8, 111
SenseFET power device 129
Servo-systems 124
Shaded-pole motors 72–3
'Skin-effect' (conductor resistivity) 26
Slip-speed 17, 153
Small motors, control 125–6
Speed control 44–7, 82, 144
  by multiple windings 85–6
  by pole modification 84–5
  by variable frequency AC source 86
  fan and blower motors 118–20
  and speed regulation 118–19
Speed control regulators (SCRs) 57–8, 118
  power factor effects 26
Speed measurement 15–17
Speed-selection system calculation 160–61
Square waveform 91–4
Starting procedures 99–100
  DC motors 99–100
  gentle 103–104
  reactance type starter 103
  three-phase induction motors 104
  use of capacitors 101–102
  wye-delta 104
Stator (definition) 10, 11
Stepping motors
  AC/DC dual role 61
  custom design 122–4
  performance improvement 122
Sub-synchronous speed operation 96–7
Switch
  BIMOS 125
  centrifugal see Centrifugal switch
Switching systems, reliability 136–7
Synchro machines 109–11
Synchro-system experiments 109–12
Synchronous speed calculations, different frequencies 150–51

Torque (turning effort) 11–12, 13
Transformer relationship, in DC motors 34
Transformer simulation calculation, induction motors 157–8
Triacs
  control circuitry 126
  power factor effects 26
Turning tendency 13

Universal motors 53, 87–8, 106

Vacuum cleaner, motor 53

Waveforms

non-sinusoidal 91–4
polyphase 108
square 91–4

Work (definition) 13